SpringerBriefs in Applied Sciences and Technology

Computational Intelligence

Series Editor

Janusz Kacprzyk

For further volumes:
http://www.springer.com/series/10618

António M. L. Canelas
Rui F. M. F. Neves
Nuno C. G. Horta

Investment Strategies Optimization Based on a SAX-GA Methodology

 Springer

António M. L. Canelas
Instituto de Telecomunicações
Instituto Superior Técnico
Lisbon
Portugal

Nuno C. G. Horta
Instituto de Telecomunicações
Instituto Superior Técnico
Lisbon
Portugal

Rui F. M. F. Neves
Instituto de Telecomunicações
Instituto Superior Técnico
Lisbon
Portugal

ISBN 978-3-642-33109-1   ISBN 978-3-642-33110-7   (eBook)
DOI 10.1007/978-3-642-33110-7
Springer Heidelberg New York Dordrecht London

Library of Congress Control Number: 2012946751

Printed on acid-free paper

Springer is part of Springer Science+Business Media (www.springer.com)

To Clara and my Parents
                        António M. L. Canelas
To Susana and Tiago
                        Rui F. M. F. Neves
To Carla, João and Tiago
                        Nuno C. G. Horta

Preface

The financial markets move vast amounts of capital around the world. This fact and the easy access to trading in a manual or automatic way that creates a more accessible way to participate in the markets activity attracted the interest of all types of investors, from the "man on the street" to academic researchers. This type of new investors and the automatic trading systems influence the market behavior. In order to adapt to this new reality, the domain of computational finance has received an increasing attention by people from both finance and computational intelligence domains.

The main driving force in the field of computational finance, with application to financial markets, is to define highly profitable and less risky trading strategies. In order to accomplish this main objective, the defined strategies must process large amounts of data which include financial markets time series, fundamental analysis data, technical analysis data, etc., and produce appropriate buy and sell signals for the selected financial market securities. What may appear, at a first glance, as an easy problem is, in fact, a huge and highly complex optimization problem, which cannot be solved analytically. Therefore, this makes the soft computing and in general the computational intelligence domains especially appropriate for addressing the problem.

The use of chart patterns is widely spread among traders as an additional tool for decision making. The Chartists, as these analysts are known, try to identify some known pattern formations and based on previous appearances try to predict future market trends. The visual pattern identification is hard and largely subject to errors, patterns in the financial time series are not as clean and high as the images in the books, so the need to create some solution that helps on this task will always be welcomed.

This work presents a new computational finance approach, combining a Symbolic Aggregate approXimation (SAX) technique together with an optimization kernel based on genetic algorithms (GA). The SAX representation is used to describe the financial time series, so that, relevant patterns can be efficiently identified. The evolutionary optimization kernel is here used to identify the most relevant patterns and generate investment rules. The SAX technique uses an alphabetic symbolic representation of data defined by adjustable parameters.

In order to capture and preserve the essence of the explored financial time series, a search for the optimal combination of SAX parameters is presented. The proposed approach considers several different chromosomes structures in order to achieve better results on the trading platform, first, it begins with a basic fixed structure and starts to evolve to an extended one that allows a more accurate use of the SAX representation, and finally finishes with a multi-chromosome structure to capture the trading potential that this approach has to offer. This approach was tested using real data from S&P500 and was also compared with another state-of-the-art approach which uses a template pattern-based method to detect patterns. The achieved results show that the proposed approach outperforms both B&H and other state-of-the-art solutions.

This book is organized into five chapters as follows:

Chapter 1 presents a brief description on the problems addressed by this book, namely the investment optimization based on pattern discovery techniques. Additionally, the main goals for the work presented in this book, as well as the document's structure are also highlighted in this chapter.

In Chap. 2, some fundamental concepts necessary to understand the developed work are addressed, particularly, the domain relative to financial markets and time series analysis. Furthermore, several methodologies applied to market investment and especially to pattern detection are presented. Finally, an introduction to the SAX representation method will be presented and previous works using this methodology will be discussed.

Chapter 3 presents an innovative methodology for pattern discovery in financial time series. The combination of SAX representation method with the use of GA to optimize the search and creation of new investment strategies will be explored. Taking advantage of the symbolic representation and dimensional reduction provided by SAX, several financial time series will be analyzed in order to search for meaningful patterns to reveal periods of time to invest on the stock market. The search and investment criteria will be defined and optimized by the use of GA, several chromosomes structures were considered in order to provide more accurate results.

Chapter 4 presents and compares the results from the approaches described in Chap. 3. The several chromosome structures were tested in real market conditions, where in all transactions the costs were considered. In order to test this new approach of investment based on pattern discovery in financial time series, two major experiences were made. The first test is based on the discovery of patterns to invest Long presented like "SAX-GA Uptrend Pattern Discovery". The second test is a method that invests Long and Short, to do this it tries to discover patterns to enter and exit Long and another set of patterns to enter and exit Short, described later on "SAX-GA Multi-Chromosome Pattern Discovery".

Chapter 5 summarizes the provided book and supplies the respective conclusions and future work.

António M. L. Canelas
Rui F. M. F. Neves
Nuno C. G. Horta

Contents

Acronyms and Abbreviations

Optimization and Computer Engineering Related

ANN	Artificial Neural Networks
DBMS	Database Management System
EA	Evolutionary Algorithm
EC	Evolutionary Computation
FSM	Finite-State Machine
GA	Genetic Algorithm
PAA	Piecewise Aggregate Approximation
PIP	Perceptually Important Points
SAX	Symbolic Aggregate approXimation
SOM	Self-Organizing Maps
SVM	Support Vector Machines

Investment Related

B&H	Buy-and-hold
CPI	Consumer Price Index
DER	Debt Equity Ratio
DY	Dividend Yield
EMA	Exponential Moving Average
EMH	Efficient Market Hypothesis
EPS	Earning Per Share
FA	Fundamental Analysis
FOREX	Foreign exchange market or currency market
FX	The same as Forex
GDP	Gross Domestic Product
HFT	High Frequency Trading
HSI	Hang Seng Index

HV	Historical Volatility
MA	Moving Average
MACD	Moving Average Convergence/Divergence
MC	Market Capitalization
OBV	On Balance Volume
PBV	Price Book Value
PCF	Price Cash Flow
PER	Price Earnings Ratio
POR	Pay Out Ratio
PSR	Price Sales Ratio
QR	Quick Ratio
ROA	Return On Assets
ROE	Return On Equity
ROI	Return On Investment
RSI	Relative Strength Index
S&P500	Standard and Poor's 500 stock index
SMA	Simple Moving Average
TA	Technical Analysis

Chapter 1
Introduction

Abstract This chapter presents a brief description on the problematic addressed by this book, namely the investment optimization based on pattern discovery techniques. Additionally, the main goals for the work presented in this book, as well as, the document's structure are, also, highlighted in this chapter.

Keywords Computational Finance · High Frequency Trading · Efficient Market Hypothesis

1.1 Financial Markets

Financial markets move vast amounts of capital around the world. This fact and the easy access to trading in a manual or automatic way that creates a more accessible way to participate in the markets activity attracted the interest of all type of investors, from the "man on the street" to academic researchers. This type of new investors and the automatic trading systems influence the market behavior. In order to adapt to this new reality the domain of computational finance has received an increasing attention by people from both finance and computational intelligence domains.

The main driving force in the field of computational finance, with application to financial markets, is to define highly profitable and less risky trading strategies. In order to accomplish this main objective, the defined strategies must process large amounts of data which include financial markets time series, fundamental analysis data, technical analysis data, and produce appropriate buy and sell signals for the selected financial market securities. What may appear, at a first glance, as an easy problem is, in fact, a huge and highly complex optimization problem, which cannot be solved analytically. Therefore, this makes the soft computing and in general the computational intelligence domains especially appropriate for addressing the problem.

The use of chart patterns is widely spread among traders as an additional tool for decision making. The Chartists, as these analysts are known, try to identify some known pattern formations and based on previous appearances try to predict future market trends. The visual pattern identification is hard and largely subject to errors, patterns in the financial time series are not as clean and thigh as the images in the books, so the need to create some solution that helps on this task will always be welcomed.

As said before, a pattern well designed as the ones that appear in the books are rare, in the real world, patterns have noise and bad formations that can cause distortions, so an automatic system to detect those kinds of patterns must be able to deal with these types of errors. Another existing problem is the fact that new patterns began to appear and quickly disappear. The market nowadays changes very rapidly thanks to the large number and behavior of investors and to the new high frequency trading systems (HFT). It was revealed by some studies [1–3] that HFT tend to stabilize markets and turn them more efficient, helping the "Efficient Market Hypothesis" (EMH) [4, 5] theory since it incorporates faster the new information, the fact is that the large number of transaction and the quick changes in volume and liquidity influence the market and creates new patterns. Although the EMH defends that it is impossible to profit by predicting the market and that HFT turn the markets more efficient, around this theory still does not exists an entire consensus of its validity. So, in this work it is studied and implemented a new method capable of detecting new and meaningful patterns that forecast future market moves and profit from them, and also beat the Buy and Hold (B&H) strategy defended by the EMH.

In this book a new approach combining a Symbolic Aggregate approXimation (SAX) technique together with an optimization kernel based on genetic algorithms (GA) is presented. The SAX representation is used to describe the financial time series, so that, relevant patterns can be efficiently identified. The evolutionary optimization kernel is here used to identify the most relevant patterns and generate investment rules.

1.2 Motivation and Work's Purpose

The most basic motivation for trying to predict stock markets is the financial gain. Several other reasons could be referred, from the engineering point a view, the challenge of elimination or reduction of uncertainty in a dynamic system like the financial sector could be a valid motivation as well. Another motivation was the fact of implementing a new solution on pattern discovery, with possible applications to other areas than the financial.

As was said before, the creation of solutions that can deal with automatic pattern discovery would be well received in the financial community. The challenge of creating a system that could deal and adapt to the fast changing world of the markets, would put to the test the fine optimization tool that is the Genetic Algorithm (GA). The possibility of developing and applying GA to this task, will allow to better understanding how it works and how well it could adapt to new challenges.

The aim of the work presented in this book, is to create an application that is based on pattern discovery and attempts to predict the stock market. To achieve this goal, the proposed approach will convert the financial data to a SAX representation, based in this symbolic representation the GA will find a set of patterns and rules, which allows investing on the stock market.

For testing this new methodology the application will invest on stocks from the S&P 500 index in real market conditions, considering transaction costs, and the results will be compared to the Buy & Hold strategy.

1.3 Book Structure

The presented book is structured as following:

- Chapter 2 addresses the theory behind the developed work, namely the concepts of Market Analysis and Investment Techniques and Soft Computing methodologies. Also, in this chapter, it is given an overview about different methodologies which can be used and are already used.
- Chapter 3 shows the new approach SAX/GA method for pattern discovery, beginning by describing the SAX method and presenting the proposed solution.
- Chapter 4 proposes the validation procedure to evaluate the developed system by providing a study of the solution's performance and robustness.
- Chapter 5 summarizes the provided report and supplies the respective conclusion and future work.

1.4 Conclusions

Algorithm trading systems caused a financial arms race, where everybody is trying to develop new and better solutions to compete on the financial markets. HFT systems changed the basic trading characteristics of the markets, these systems are able to take decisions and analyze the market faster than the traditional human broker. This kind of trading systems has a large effect on the market and creates new patterns on the stock charts. The ability of detecting these new formations could improve the capability of forecasting the financial markets. The new approach, presented in this book, is capable of detecting new and meaningful pattern formations in order to assist on a trading decision system.

References

1. V.H. Martinez, I. Rosu, High frequency traders, news and volatility (2011), http://appli8. hec.fr/rosu/research/news.pdf
2. J. Brogaard, High frequency trading and volatility (2012), http://dx.doi.org/10.2139/ssrn. 1641387
3. J. Brogaard, High frequency trading and market quality (2010), http://dx.doi.org/10.2139/ssrn. 1970072
4. B.G. Malkiel, The efficient market hypothesis and its critics. J. Econ. Persp. **17**(1), 59–82 (Winter 2003). DOI:10.1257/089533003321164958
5. B.G. Malkiel, *A random walk down Wall Street* (W. W. Norton & Company, New York, London, 1999)

Chapter 2
Market Analysis Background and Related Work

Abstract In this chapter some fundamental concepts, necessary to understand the developed work, are addressed, particularly the domain relative to financial markets and time series analysis. Furthermore several methodologies applied to market investment and especially to pattern detection are presented. Finally an introduction to the SAX representation method will be presented and previous works using this methodology will be discussed.

Keywords Technical Analysis · Fundamental Analysis · Perceptually Important Points (PIP) · Symbolic Aggregate Approximation (SAX) · Pattern Recognition · Pattern Discovery

2.1 Market Analysis

In order to understand the objective of this work, it is necessary to have some basic knowledge of market analysis. Basically, there are two market analysis techniques, the *Fundamental Analysis* and the *Technical Analysis*, these are presented next. For the present work and almost for all automatic investment algorithms the preferred analysis method is the technical, since it is based on some measurable numeric indicators, easily calculated from stock market time series. Another reason for using of this type analysis is the fact that data is easily obtained from the internet, most of the time for free, and the large of number of data available to train and test the algorithms.

A. M. L. Canelas et al., *Investment Strategies Optimization Based on a SAX-GA Methodology*, SpringerBriefs in Computational Intelligence, DOI: 10.1007/978-3-642-33110-7_2, © The Author(s) 2013

2.1.1 Fundamental Analysis

The fundamental analysis (FA) is based on several economic and financial indicators, which attempt to evaluate the intrinsic value of a company. FA studies everything that can affect the company value, including macroeconomic and business specific factors. So, the fundamental analysts will try to determine the future price of a company and based on the current value takes the investment decision.

Some of the most important economic indicators are released by central banks or other institutions, and include indicators like:

- Unemployment
- Consumer Price Index (CPI)
- Consumer Confidence Index
- Gross Domestic Product (GDP)
- New Home Sales

Other indicators could be found in [1], where an important and detail list of indicators are presented.

The fundamental analysts also use indicators that are related to company analysis. These are gathered from the financial reports issued by companies. Some of the most important ones are presented in Table 2.1.

Finally, industry reports are the source to another important set of indicators. With these industry indicators the analysts try to evaluate how healthy is some particular sector, and how the companies are positioned inside it. One important factor to every investor is how to minimize the risk; one way to achieve this goal is to investment on different industries to avoid possible down cycles of some sectors. Some important indicators to look out are:

- Industry Growth
- Competition
- Costumers
- Suppliers

Since each industry has some business specificity is more difficult to make a detail list of indicators.

2.1.2 Technical Analysis

Technical analysis [2] is based on the stock prices and volumes movements, the technical analyst believe that changes on price and volume already incorporates all the fundamentals factors. In this technique, the stock price and volume is all that matters to describe the market condition and to try to predict future market movements. Based on this premises, the analyst builds a set of indicators that allows him to study, in an easier and deeper way, the stock movements in order to profit from future trends.

Table 2.1 Several important company fundamental indicators

Indicator	Measure	Description
Earning per share (EPS)	$\dfrac{\text{Net Income} - \text{Dividends on preferred Stock}}{\text{Average Outstanding Shares}}$	The portion of a company's profit allocated to each outstanding share of common stock
Price earning ratio (PER)	$\dfrac{\text{Market Value per Share}}{\text{EPS}}$	A valuation ratio of a company's current share price compared to its per-share earnings
Price cash flow (PCF)	$\dfrac{\text{Share Price}}{\text{Cash Flow per Share}}$	A measure of the market's expectations of a firm's future financial health
Price sales ratio (PSR)	$\dfrac{\text{Share Price}}{\text{Reveneu per Share}}$	A ratio for valuing a stock relative to its own past performance
Payout Ratio (POR)	$\dfrac{\text{Dividends per Share}}{\text{EPS}}$	The amount of earnings paid out in dividends to shareholders
Dividend yield (DY)	$\dfrac{\text{Anual Dividends per Share}}{\text{Price per Share}}$	A financial ratio that shows how much a company pays out in dividends each year relative to its share price
Price book value (PBV)	$\dfrac{\text{Stock Price}}{\text{Total Assets} - \text{Intangible Assets} - \text{Liabilities}}$	A ratio used to compare a stock's market value to its book value
Return on equity (ROE)	$\dfrac{\text{Net Income}}{\text{Shareholder's Equity}}$	The amount of net income returned as a percentage of shareholders equity
Return on assets (ROA)	$\dfrac{\text{Net Income}}{\text{Total Assets}}$	An indicator of how profitable a company is relative to its total assets
Debt equity ratio (DER)	$\dfrac{\text{Total Liabilities}}{\text{Shareholders Equity}}$	Measure of financial leverage
Quick ratio (QR)	$\dfrac{\text{Current Assets} - \text{Inventories}}{\text{Current Liabilities}}$	Short-term liquidity, ability to meet short-term obligations
Market capitalization (MC)	Company's Shares $\times$ Market Price	The total market value of all of a company's outstanding shares

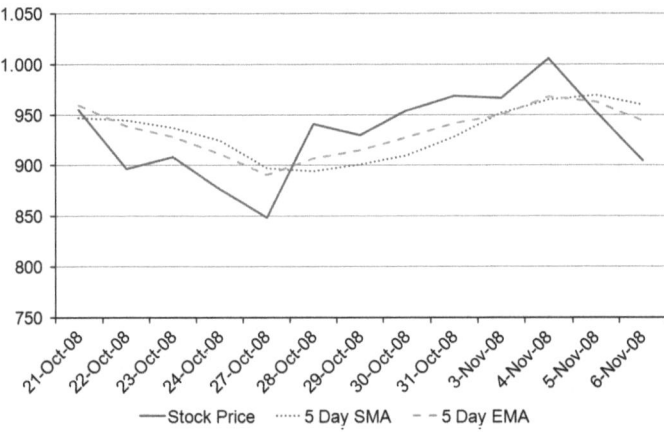

Fig. 2.1 S&P 500 daily chart with a SMA and an EMA

In addition to the technical indicators, that are calculated using the price and transaction volumes, the technical analyst studies pattern formations [3, 4] in the stock price and sometimes in some indicators.

Next, some technical indicators and relevant known pattern formations are presented.

2.1.2.1 Technical Indicators

A technical indicator is a metrics whose value is calculated from the price/volume of an asset. The objective is that the indicator value helps predicting future price, or simply indicates a general price trend. Some popular technical indicators are presented next, for more information on other indicators see [5].

- Moving Averages

This is one of oldest indicator used and is calculated by finding the mean value of the price over a certain amount of time. Two type of moving average are applied, the first one is known as Simple Moving Average (SMA or just MA) and is calculated by averaging the price of the last days, or any other time measure. The second average is the Exponential Moving Average (EMA) and in this case the price for recent days has a higher weight on the average.

In Fig. 2.1 the S&P500 index is presented, in this figure two averages are shown, the SMA for 5 days period and the EMA for the same period. It is clear the difference between them, the EMA follows better the fast changes of price indicating that the present time has more weight.

This indicator could be used to signal the investor to buy when the moving average is rising or crosses down the price line and sell when is descendant or crosses up the price line. Sometimes the moving averages are used together to

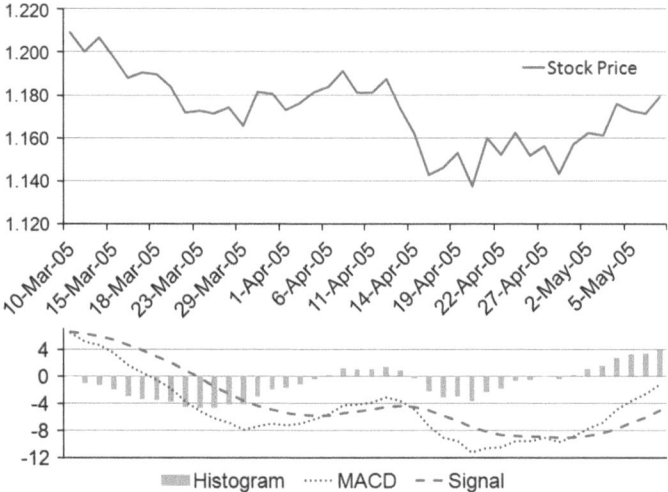

Fig. 2.2 S&P 500 daily chart and MACD indicator, histogram and signal line

generate these signals, the intersection between different moving average is used to generate buy or sell orders.

- MACD (Moving Average Convergence Divergence)

One example of the combined used of averages is this indicator, which is calculated from the difference of two EMA, usually the 12 days EMA and the 26 days. An additional line is added to this indicator, is known as the Signal line and is the 9 days moving average of the MACD itself. Added to this indicator usually appears a histogram, which is the difference between the MACD and the Signal, Fig. 2.2. Sometimes the MACD line does not appear, instead just the signal line and the histogram are presented, this is because usually the cross point of the two lines is the point to look for and that point could be identified by the zero crossing of the histogram.

The application rules to the MACD indicator are discussed next:

- The crossing between MACD and the Signal when the MACD is rising is indication to buy in a down movement is to sell.
- The zero cross by the MACD is signal to buy, since when MACD is above zero the market tends to be bullish.
- Positive or negative divergence is also a signal. If the price is on uptrend and the MACD does not, then a negative divergence is present and is time to sell. If the price in downtrend and the MACD begins an uptrend then is a buy signal.

- RSI (Relative Strength Index)

Fig. 2.3 RSI indicator for the S&P 500 daily chart

This indicator is one of the most popular in the momentum family, which compares the magnitude of recent gains to recent losses to identify overbought and oversold conditions of an asset. This indicator representation oscillates between 0 and 100 and usually is calculated for 14 days period. In Eq. (2.1) the formula for calculating this indicator is presented (Fig. 2.3).

$$RSI = 100 - \frac{100}{1 + \frac{\bar{G}}{\bar{L}}} \qquad (2.1)$$

where,

$\bar{G}$ Average gain of the x last periods of time;

$\bar{L}$ Average loss of the x last periods of time.

In the use of this indicator it is also possible to detect graphical pattern formations, or by applying the following rules:

- Crossing the 50 value in an uptrend is a buy signal, in case of a downtrend indicates a sell signal.
- If the RSI is above 70 is an overbought signal and the asset should be sold. Below 30 the asset is oversell and a buy signal is issued.
- Like in MACD, this indicator presents positive and negative divergence. If the price is in a downtrend and the RSI does not follows then a buy signal must be

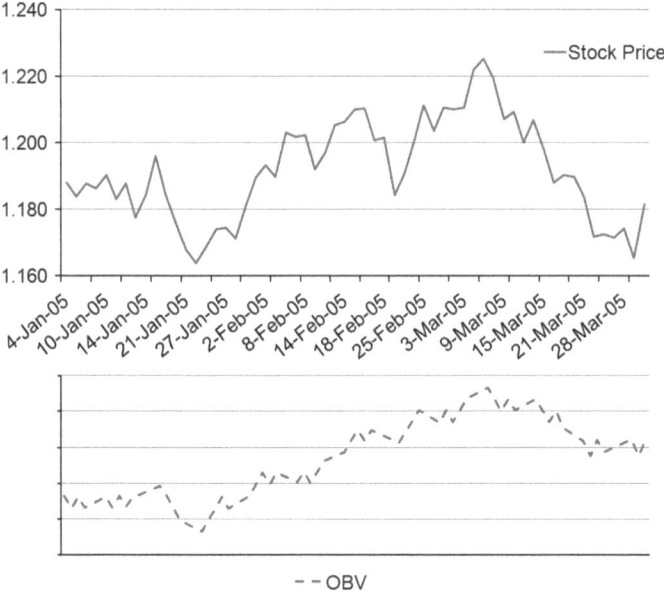

Fig. 2.4 OBV indicator for the EUR/USD Forex pair

issued, this is a positive divergence. If the price is at maximums and the RSI is not, then a negative divergence is present and the asset should be sold.

- OBV (On Balance Volume)

This indicator, as the name states, uses the volume for its calculation and the price as well. It measures buying and selling pressures. It supports the idea that volume precedes price, the value itself is not important what is important is characteristics of the OBV line. The OBV is calculated by Eq. (2.2) (Fig. 2.4).

$$OBV(t) = \begin{cases} OBV(t-1) + Volume(t), & if\ Price_t > Price_{t-1} \\ OBV(t-1) - Volume(t), & if\ Price_t < Price_{t-1} \\ OBV(t-1), & Otherwise \end{cases} \quad (2.2)$$

Several signals could be perceived from this indicator:

- A bullish market is in formation when OBV moves up or forms a higher low, even if the prices move down or forms a lower low, in this case a buying signal is issued. A bearish market forms when OBV moves down or forms a lower low, even when the prices move up or a higher high is created, a sell signal is generated for this case.
- In case the OBV slope is positive and the price is in an uptrend this is a confirmation of the trend. The same applies for the opposite case.

Fig. 2.5 Head and shoulders

Fig. 2.6 Inverse head and shoulders

Fig. 2.7 Symmetrical triangle in a uptrend

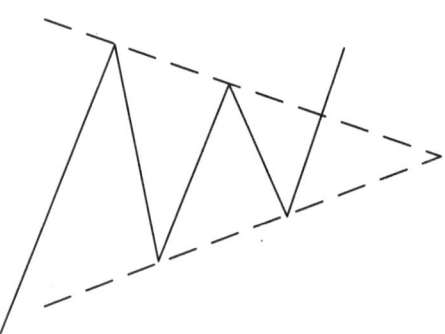

2.1.2.2 Chart Patterns

When looking at a graphic representation of a financial time series, it is possible to identify some similar graphic formation along the time. These formations are caused by the repeated actions of the investors when presented to similar market conditions. So, by looking for these similar graphic formations or chart patterns, it will be possible with some degree of confidence to deduce what will happen next. This is the base idea behind the technical analysts that hunt for chart patterns and are known as Chartists.

Next, a set of important chart patterns is presented, the associated investment strategy is easily guessed by the outcome of the patterns. The presented patterns are shown due to its relevance and in order to possible identify some of the patterns found by the SAX-GA approach. Many more patterns exists, for more information on this graphic formations see [3], and for additional trading tips using patterns [4] and more specific to the Foreign Exchange Market (Forex) [6].

Fig. 2.8 Symmetrical
triangle in a downtrend

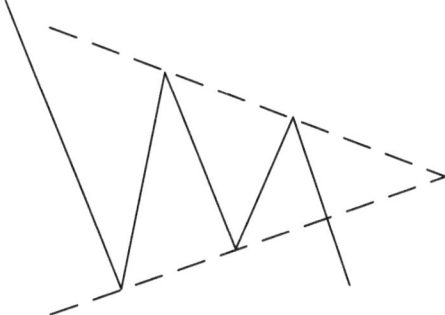

Fig. 2.9 Ascending triangle

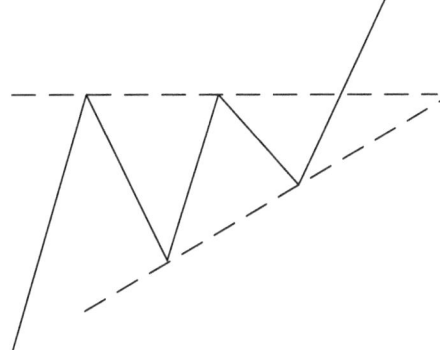

Fig. 2.10 Descending
triangle

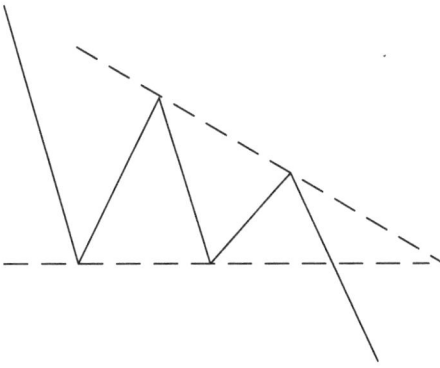

• Head and Shoulders

   This is one of the most famous chart formations. This is a reversal type that
indicates a change on trend. The two possible reversal trend formations are presented in Fig. 2.5 and Fig. 2.6.

Fig. 2.11 Falling Wedge in an uptrend

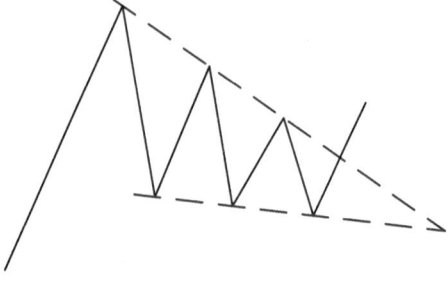

Fig. 2.12 Falling wedge in a downtrend

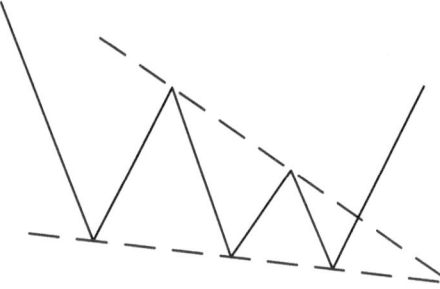

Fig. 2.13 Rising wedge in an uptrend

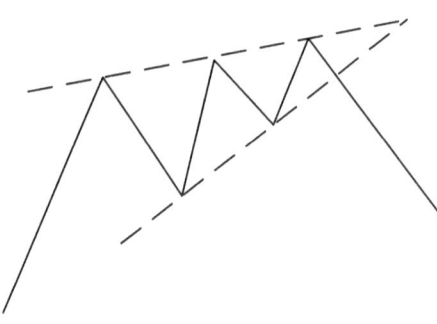

Fig. 2.14 Rising Wedge in a downtrend

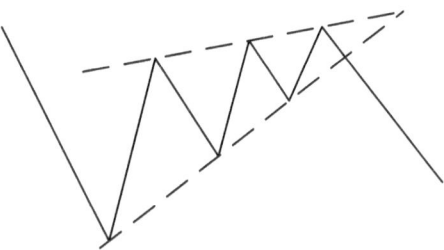

- Symmetrical Triangles

   This formation is usually defined as neutral, the fact is that some researchers found that most of the time breaks in the previous market direction, could be considered as a trend confirmation. This is illustrated in Fig. 2.7 and Fig. 2.8.

Fig. 2.15 Bull flag in an uptrend

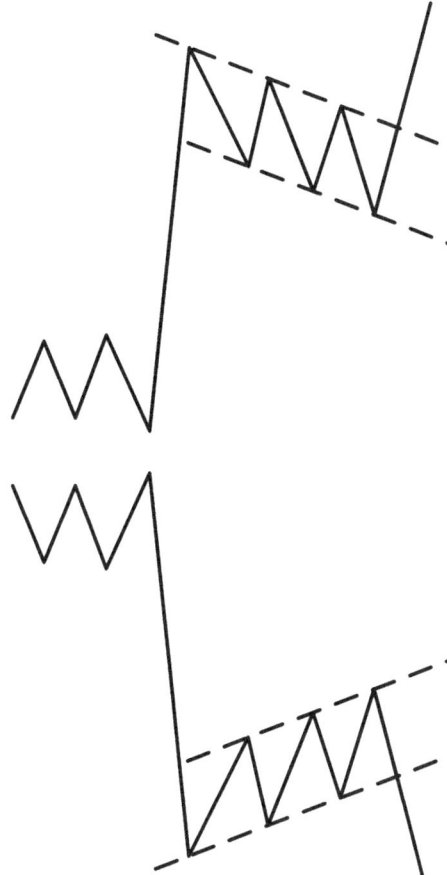

Fig. 2.16 Bear flag in a downtrend

- Ascending Triangle/Descending Triangle

  Ascending triangles are considered an uptrend confirmation and naturally are most reliable when found in bullish market conditions, Fig. 2.9.

  The descending formation, Fig. 2.10, is the reverse of the previous one, is also a trend confirmation, and naturally is more reliable in a bearish market.

- Falling Wedges

  These formations, Figs. 2.11 and 2.12, are quite similar to the triangles, but they are associated to a bullish movement, breaking in an uptrend.

- Rising Wedges

  Rising Wedges are the inverse of the previous chart formation and are associated to bearish movements, Fig. 2.13, and appear more frequently in a downtrend market, Fig. 2.14.

Fig. 2.17 Pennant in an
uptrend

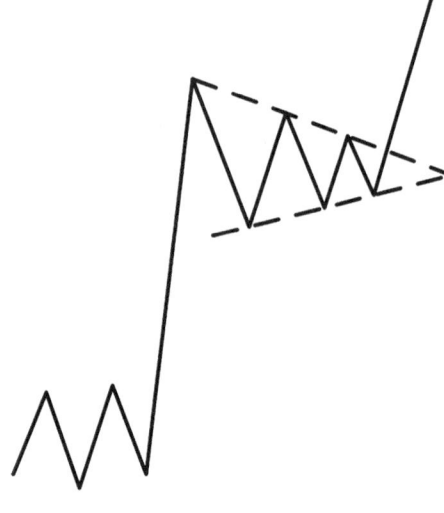

Fig. 2.18 Pennant in a
downtrend

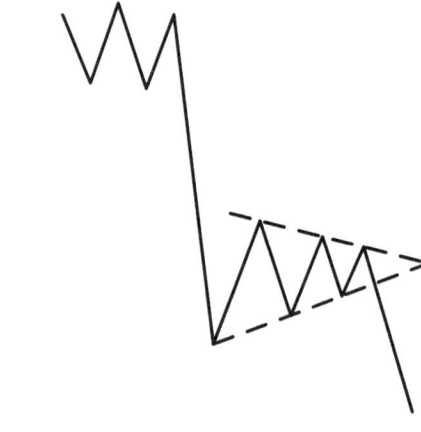

Fig. 2.19 Triple top

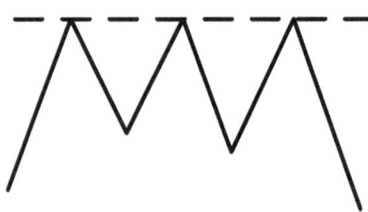

• Flags

These charts formations are generally seen as trend confirmations, usually
appear after big movements on the graphic, as in Fig. 2.15 and 2.16

Fig. 2.20 Double bottom

- Pennants

These are similar to Symmetrical Triangles, but usually are smaller in amplitude and shorter in time, Figs. 2.17 and 2.18. Like the Flags, the Pennants are generally trend confirmations and represent small pauses in the current trend.

- Tops and Bottoms

These formations are usually trend reversals, they represent a price resistance or a support level, Figs. 2.19 and 2.20 respectively.

2.2 Existing Solutions

In this section several state-of-the-art time series representation and dimensional reduction methods, applied to the financial sector, are presented. Also, their application and use on chart pattern detection methods will be discussed, and at last the SAX method will be briefly presented.

2.2.1 Pattern Detection

From the beginning of stock market history, investors tried to predict market movements. The analysis has always been difficult, and one of the harder tasks is to analyze financial charts, in order to detect patterns on those graphics. The appearance of computers and their large calculus capabilities offers new possibilities of prediction.

First of all a distinction between pattern recognition and pattern discovery should be made. Recognition is identifying some patterns that are known on the time series, this case is a supervised approach, where a library of patterns [3], is created and is made a search on the data market trying to identify them [7]. In pattern discovery, the quest is to find patterns that occur in the time series and that are unknown, in this case typically some data segments or windows are compared with others and this case is associated to unsupervised approach, the case presented on this book.

The methods that directly compete with the new SAX-GA approach, are techniques that try to detect chart patterns from the financial time series. Basically, are alternative representations of data series and then apply some decision or

Fig. 2.21 PIP identification process

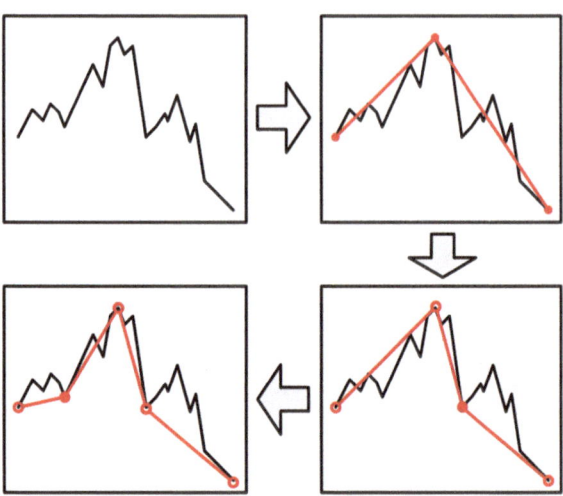

classifier method to identify the patterns. In this area two methods were identified, the first one uses a Perceptually Important Points (PIP) representation and the second uses a matrix representation of data. The matrix data representation and PIP when are used in a pattern detection methods usually are applied in a template based pattern recognition strategy. Where a set of known patterns, are converted to the same representation structure and a matching process, between template and time series, is made. Generally, this matching process is based on the distance measure between the data representation and the template, and if the distance is lower than some threshold, it is considered that the pattern is present in the evaluated time series. A detailed survey comparing several time series data mining techniques is presented in [8].

2.2.1.1 Perceptually Important Points Pattern Recognition

As the name of this alternative data representation and dimensional reduction indicates, this method reduces the time series to a set of points that are considered important. The decision to identify the important points is quite similar to the one a person makes when looking at a graphic, where the points that more stand out from each other are the ones that attract the eye attention. This pattern recognition method was first present by Chung et al. [9] and it first use was in financial applications.

The process to identify PIP's in a time series is quite simple and could be seen in Fig. 2.21. It starts by defining that the first two important points are the first and the last of the time series, then draws a line between those points and calculates for the rest of the points in the time series which is further apart from the line, the more distant point will now be an important point. After, draws other line between the first point and this new one, and from the new to the last point, and for each of

Fig. 2.22 PIP evaluation
with Euclidian distance
(PIP-ED)

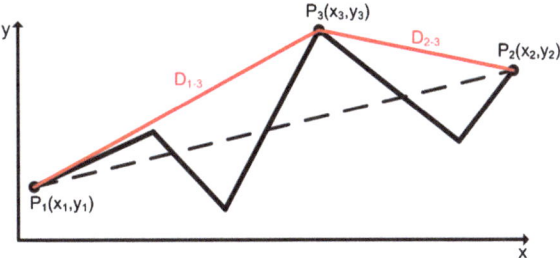

Fig. 2.23 PIP evaluation
using perpendicular distance
(PIP-PD)

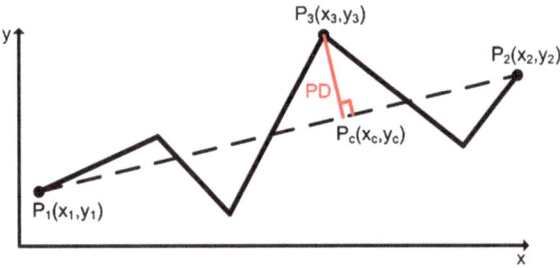

the segments, again finds the point further apart, and in a recursive way it will
detect the important points [10]. If this process is taken to the limit then all points
from the time series will be important points. So, when to stop? Well, there is no
straight answer to this question since is data dependable, if the time series is very
irregular with lots of oscillations then more PIP's have to be found to preserve the
data structure. Usually is defined a desired compression ratio, Eq. (2.3), or an
acceptable level of error between time series and the PIP representation.

$$C_r = \frac{Number\ of\ points\ in\ time\ series}{Number\ of\ PIP's\ to\ represent\ time\ series} \tag{2.3}$$

In the method description, was said that the distant point from the line will be
considered an important one. So it is important to define how to measure the
distance, in [10] three types of measures were considered and are presented next:

- Euclidian distance (ED)—Calculates the sum of distances, Eq. (2.4), between
  the test point p_3 and the adjacent PIP's p_1 and p_2—Fig. 2.22

$$ED(p_1, p_2, p_3) = \sqrt{(x_2 - x_3)^2 + (y_2 - y_3)^2} + \sqrt{(x_1 - x_3)^2 + (y_1 - y_3)^2} \tag{2.4}$$

- Perpendicular distance (PD)—Calculates the PD, Eqs. (2.5)–(2.8) between the
  test point p_3 and the line that connects the two adjacent important points p_1 and
  p_2 —Fig. 2.23

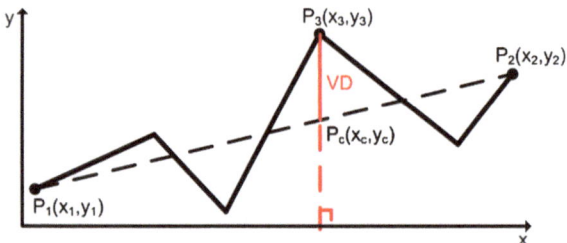

Fig. 2.24 PIP evaluation
using vertical distance
(PIP-VD)

$$s = Slope(p_1, p_2) = \frac{y_2 - y_1}{x_2 - x_1} \tag{2.5}$$

$$x_c = \frac{x_3 + sy_3 + sy_2 - s^2 x_2}{1 + s^2} \tag{2.6}$$

$$y_c = sx_c - sx_2 + y_2 \tag{2.7}$$

$$PD(p_3, p_c) = \sqrt{(x_c - x_3)^2 + (y_c - y_3)^2} \tag{2.8}$$

- Vertical distance (VD)—Calculates the VD, Eqs. (2.9)–(2.10), between the test point p_3 and the line that connects the two adjacent important points p_1 and p_2—Fig. 2.24

$$y_c = y_1 + (y_2 - y_1) \frac{x_3 - x_1}{x_2 - x_1} \tag{2.9}$$

$$VD(p_3, p_c) = |y_c - y_3| \tag{2.10}$$

From tests made using the Hang Seng Index (HSI), the distance method that proves best results is the vertical distance, which was able to capture the essence of the HSI graphic [10].

Now, to complete the pattern recognition method description is necessary to compare the converted time series with the templates, for instance Fig. 2.25.

So, if P important point's exists in the sequence, converted from a time series, and a query template sequence Q is defined, is possible to calculate the distance point-to-point from each other, Eq. (2.11).

$$AD(P, Q) = \sqrt{\frac{1}{n} \sum_{k=1}^{n} (p_k - q_k)^2} \tag{2.11}$$

where,
p and q Points of the sequence;
n Number of points in the sequence.

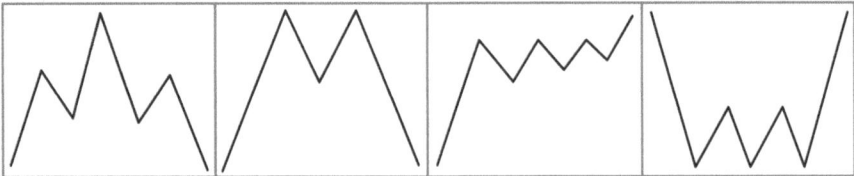

Fig. 2.25 Four PIP technical analysis patterns templates

Fig. 2.26 Matching between template and time series

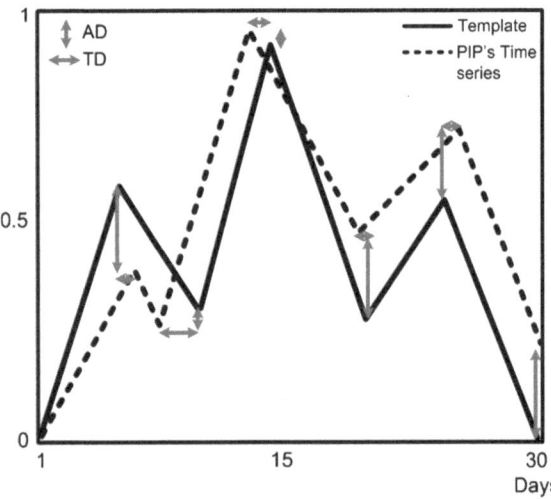

This last distance measure, Eq. (2.11), makes possible to compare the sequences according to the amplitude and in this manner identify the amplitude similitude of the sequences, but it is also necessary to detect the similarity in the horizontal temporal axes, it is also important to consider time distortions between time series and template, so it is necessary to calculate a temporal distance Eq. (2.12).

$$TD(P,Q) = \sqrt{\frac{1}{n-1} \sum_{k=2}^{n} (p_k - q_k)^2} \tag{2.12}$$

In this measure, Eq. (2.12), the sum begins at 2 because the first points of the two sequences are aligned, so the distance is zero. The p and q are the time coordinates of the sequences.

Now that a set of measures that provides a way to compare the time series and the template have been defined, Fig. 2.26, it is possible to combine the two distances in one expression to measure the similarity between the two Eq. (2.13), and to be able to decide if the pattern is present in the tested data series or not.

$$DM(P,Q) = wAD(P,Q) + (1+w)TD(P,Q) \tag{2.13}$$

where,

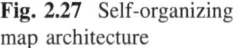

Fig. 2.27 Self-organizing
map architecture

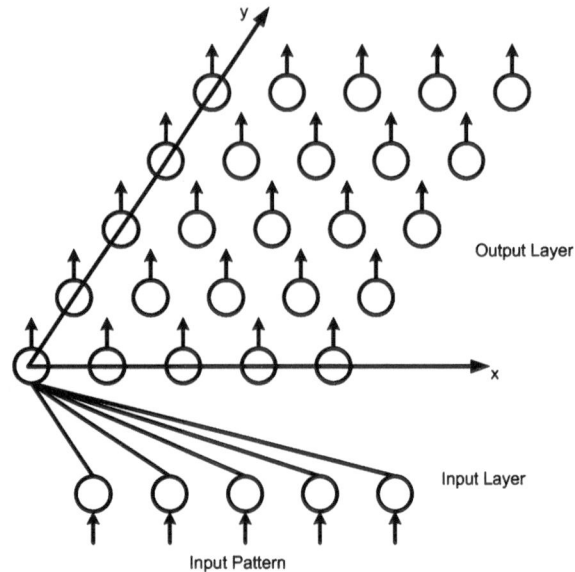

w Weight factor that allows to specify which measure is more important,
according to tests made in [11], 0.4 is a reasonable value for this factor.

The time series representation using PIP's, has the advantage of preserving
some of the important features of financial time series. Many of the preserved
points are important indicators of trend inversion. One of the problems when using
this method to recognize patterns, is that it is necessary to convert the time series
to the same number of PIPs present in the template, in order to evaluate the degree
of similitude between time series and template. So, if the application has to find a
head-and-shoulders pattern, it is necessary to convert the time series to a seven
important points representation. After, if template changes to a different pattern
with a different number of important points, it is necessary to reprocess the time
series to match the same number of points. The authors tried to solve this problem
by loading the PIP's to a binary tree like structure, where the most important points
are stored in the higher levels of the tree and the less important are near the leaves
[10]. This way it is easy to prune several braches of the tree to have the needed
number of PIP's representation.

In [12] the authors also feel the need of pattern discovery, of creating a system
where the data series itself define the patterns. To solve this problem the authors
used a Self-Organizing Map (SOM) which is a neural network based unsupervised
learning algorithm that allows similar time series data windows to be clustered
together. SOM is based on competitive learning where the only one of the output
nodes is activated. The winner node will then see their weights adjusted this will
cause that this node becomes specialized on the kind of patterns that cause the
activation. The SOM structure used is presented in Fig. 2.27.

Fig. 2.28 Bull flag matrix
pattern template

0.5	0	-1	-1	-1	-1	-1	-1	-1	0
1	0.5	0	-0.5	-1	-1	-1	-1	0.5	0
1	1	0.5	0	-0.5	-0.5	-0.5	-0.5	0	0.5
0.5	1	1	0.5	0	-0.5	-0.5	-0.5	0	1
0	0.5	1	1	0.5	0	0	0	0.5	1
0	0	0.5	1	1	0.5	0	0	1	1
-0.5	0	0	0.5	1	1	0.5	0.5	1	1
-0.5	-1	0	0	0.5	1	1	1	1	0
-1	-1	-1	-0.5	0	0.5	1	1	0	-2
-1	-1	-1	-1	-1	0	0.5	0.5	-2	-2.5

The authors identified two major problems; the first was the efficiency of the discovery process, the increase of the number of data points in the pattern lead to an exponential increase of the pattern discovery. The second was the multi-resolution problem, where the patterns can appear with different lengths, causing to reprocess the time series with different SOM architectures. In the paper, the authors solve the problems by using the PIP representation and converting a set of different lengths data window time series, to the same number of PIP's. This will limit the number of points in the patterns and also convert the data to the same representation, allowing to use the same SOM architecture. This approach will cause, that in some cases, the compression of data will be rather large and some important features will be lost, also the training process is long, needing many iterations trough the training set.

From the previous descriptions of this approach, it is clear that the PIP method has lots of potentials, from the representation point of view and pattern recognition, but creating an algorithm to discover new pattern formations using this method, will result on a time consuming approach. The multi-resolution problem can affect the pattern discovery process and increasing the compression of data is not the solution, since will cause the loss of important information. To solve this problem, the algorithm will have to save the several dimensional representations of the time series windows and then search between the windows with equal number of PIP's, causing a large and complex process of pattern search.

2.2.1.2 Matrix Template Pattern Recognition

This approach is based on the works of Leigh et al. [13–15]. As the name implies this method will recognize patterns based on a template pattern approach and the templates are in a matrix format, like Fig. 2.28. Is visible by the matrix values, that

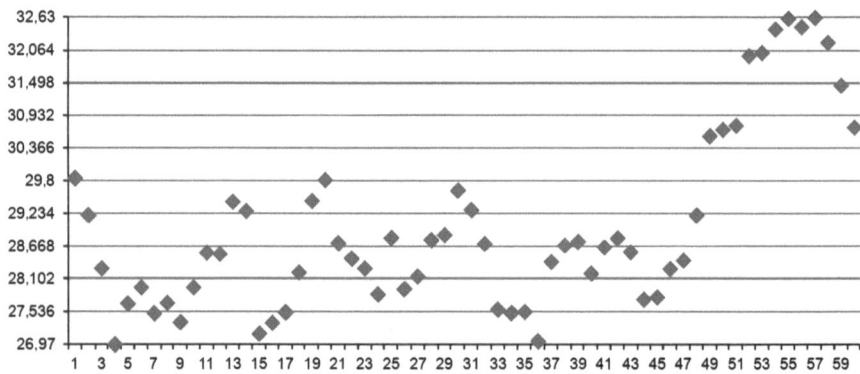

Fig. 2.29 60 points time series

Fig. 2.30 Matrix
representation of the time
series

0	0	0	0	0	0	0	0	0.17	0.67
0	0	0	0	0	0	0	0	0.33	0
0	0	0	0	0	0	0	0	0	0.17
0	0	0	0	0	0	0	0	0.5	0..17
0.17	0	0	0.17	0	0	0	0	0	0
0	0	0.33	0.17	0	0.17	0	0	0	0
0.17	0	0	0.17	0.5	0.17	0.5	0.17	0	0
0.17	0.33	0.17	0.33	0.17	0	0.5	0.5	0	0
0.33	0.33	0	0.17	0.17	0.33	0	0.33	0	0
0.17	0.33	0.5	0	0	0.33	0	0	0	0

the region where the pattern is present is populated by the maximum value 1, other
regions of the matrix farther away from the pattern, have negative values. So, the
method used to match the time series with the template, is based on the conversion
of the time series to an identical size matrix, like the template, and the result of the
product between the two matrixes, will indicate the level of similarity between
time series and pattern.

The conversion of the time series to a matrix format, consists on dividing the
time series or a window of the time series, in a grid of 10×10, for the present
example. To convert the time series like the example in Fig. 2.29, the vertical axes
will be divided in 10 levels and on the horizontal temporal axes the 60 points will
be divided into 10 groups of 6 points each. Accordingly to the number of points
that lay on each cell of the grid a percentage value is calculated, with the constraint
that the sum of these values will be 1 for each column, Fig. 2.30.

So, after converting the times series to a matrix format, a match between template and data must be calculated, and depending on the similarity result an investment decision should be made.

The authors have tested an artificial neural networks (ANN) and genetic algorithm (GA) to implement an investment decision model [16, 17]. The ANN will have as inputs the sum of the product between template and price/volume matrix by column, so from these operation the ANN will have 20 inputs, an additional 2 input values will be considered, those values correspond to the window height of price and volume, these parameters correspond to the relation of the difference between the lowest price/volume and the highest price/volume with the price/volume window size. The output of the ANN will be a price forecasting in one of the case studies and in the other, the two outputs contain a confidence factor and based on a threshold mechanism tries to predict the market. The GA was used by one of these approaches [17] to reduce the input number of variables to the ANN by determining a sub set of the 22 inputs that optimize the R2 correlation coefficient between the neural estimated price increase and the actual. The results are shown in Table 2.3 at the final of the chapter.

This method has several limitations, when the patterns to be recognized are more complex, for instance a head-and-shoulders, since the level of detail that the matrix has to offer is not enough to represent those kind of complex patterns. It is possible to increase the matrix size, but then the time series window size will also have to increase. This approach is entirely design to work as pattern recognition method with simple patterns. Although, the use of the GA proves to get good results, this algorithm was able to improve the performance of the ANN, by selecting a set of important parameters to feed the ANN classifier. In [7] the GA was also used, with good results, but in this case to select the degree of similitude between template and data in order to create trading rules.

2.2.1.3 Computational Intelligence

This area of computational science and specially when connect to the finance sector had suffer a rapidly expansion. Evolutionary computation (EC) like the GA has been largely applied, like in the examples of the previous section. Other works, instead of dealing with patterns, have used GA to find sets of technical indicators and tune their parameters [18] in order to define profitable trading rules, in this work was also addressed the problem of overfitting, which this kind of algorithms tend to suffer. A similar work [19], defines trading rules based on a set of MA chosen by the GA, in order to define an investment decision support system.

Other machine learning methodologies had been applied to the financial sector. ANN has been used with success in classifying market condition and forecast future prices of assets [20]. Other method largely used is the Support Vector Machine (SVM), which has been used as a classifier and as an estimator regression [21, 22] in order to predict future market conditions.

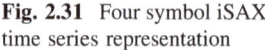 **Fig. 2.31** Four symbol iSAX
time series representation

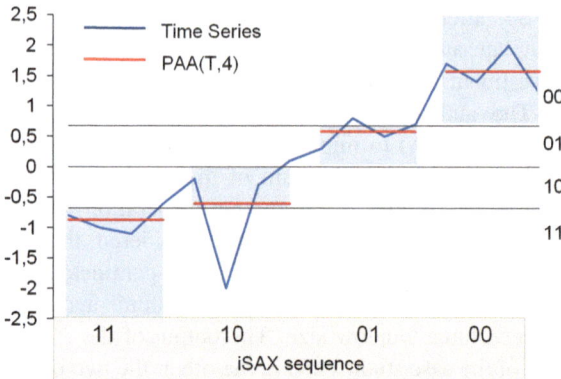

The fact is that, in several works, these last methods always have been used as classifiers and optimization methods to predict markets, based on technical indicators or to recognize patterns and never to detect new pattern formations. The use of clusters to discover patterns, like in [12], prove that this kind of solutions did not offer the flexibility that allows the algorithm to adapt to the data and patterns characteristics. The ability that GA reveals, by the easy way how a problem could be coded in their genes, in order to optimize parameters and find solutions, also the possibility of creating chromosome structures that could change dynamically, makes the GA the ideal candidate to discover new pattern formations.

2.2.1.4 Symbolic Aggregate ApproXimation

In these last few years the appearance of algorithms for efficient string manipulation and the bioinformatics applications, for instance the Human Genome Project (HGP), turn the scientific community attention to the use of symbolic representation of data. In 2003 this symbolic method was first presented by Lin et al. [23, 24].

Traditionally time series representation and data dimensionality reduction use numeric methods, like Discrete Fourier Transform (DFT) [25] or Singular Value Decomposition (SVD) [26], those methods allow to define a similarity metrics between data representation that relates to the real distance between the raw series. The use of symbolic representation of time series had always suffer from the fact that the distance between sequences have low correlations to the distance defined between the original time series. The SAX method solves the problem of distance between representation and real data, since a lower bounding approximation from the Euclidian distance could be obtained [27].

Symbolic aggregate approXimation (SAX) is based on the Piecewise Aggregate Approximation (PAA) [26], which basically consists in dividing the time series in equal size segments and then calculates the mean of the points in each segment, this new value will represent that segment. With this new representation a dimensional reduction of the time series is possible. To finally get to SAX, the

Fig. 2.32 Two symbol iSAX
time series representation

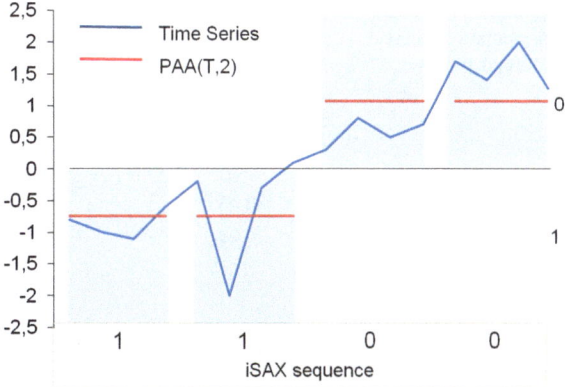

PAA method suffers an additional step at the end, where the mean value for representing the time series section is discretized to a symbol, this process will be presented on Chap. 3.

One of the advantages of using a symbolic representation, when using a database management system (DBMS) to hold the time series representation, has to do with the fact that the index structure of the DBMS can easily index sequences of strings and allows to query data to search for patterns. To improve time series data mining and indexing, when dealing with large amounts of data, an extension of the standard SAX was implemented [28], this new method is "indexable Symbolic Aggregate approXimation" or iSAX. This new approach allows for a fast time series data mining and reduced index timing, in [29], using iSAX 2.0, was possible to reduce the index building time of a data series by 72 % when comparing to standard iSAX, for this tests were used 1,000,000,000 (one billion) time series of length 256 and took less than 400 h. This new iSAX representation replaces the alphabetic symbols by binary sequences, Fig. 2.31.

The special codification of binary sequences allows creating less detail representations for the same data, with less number of symbols. For instance in Fig. 2.32 is the same time series of the previous Fig. 2.31, where it is possible to verify that thanks to the careful choice of the symbolic binary representation the symbols below zero all begin with the number one and above with zero, on both figures. So if the system codifies the time series with greater detail it is possible to go to a rough representation by removing trailing bits.

This characteristic allows the creation of an index tree structure that allows to fast search for a specific representation [28]. The present work did not use this kind of representation since the SAX-GA approach tries to discover new meaningful patterns in a time series and not search known patterns in time series database.

Another important variation of SAX for financial applications is eSAX, which stands for extended SAX [30]. This new approach to the SAX method tries to avoid the loss of some important characteristics of the financial time series by adding additional information to the representation. As was identified by studies using PIP, some important points exists in financial data, those points could

Fig. 2.33 PAA/SAX
representation misses
important points

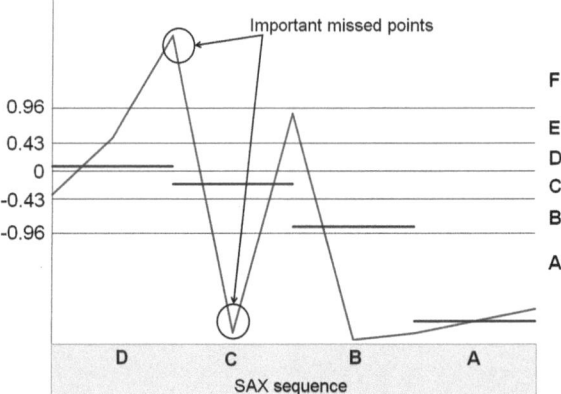

Fig. 2.34 eSAX time series
representation

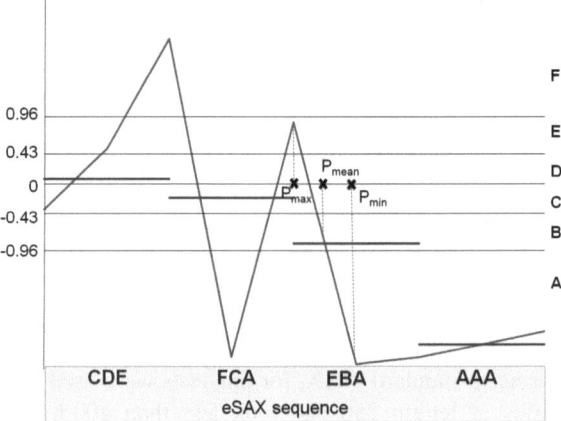

indicate a reverse trend on the market and the standard SAX tends to smooth them
since is based on PAA, which is a method based on mean values, Fig. 2.33.

As can be seen on Fig. 2.33 some important points that could increase volatility
are missed by the SAX representation, to solve this problem eSAX uses three
symbols per division of the time series indicating the maximum, mean and minimum points in each division, so instead of one symbol per segment this method
uses three symbols according to Eq. (2.14).

$$\langle S_1, S_2, S_3 \rangle = \begin{cases} \langle S_{max}, S_{mean}, S_{min} \rangle, & \text{if } P_{max} < P_{mean} < P_{min} \\ \langle S_{min}, S_{mean}, S_{max} \rangle, & \text{if } P_{min} < P_{mean} < P_{max} \\ \langle S_{min}, S_{max}, S_{mean} \rangle, & \text{if } P_{min} < P_{max} < P_{mean} \\ \langle S_{max}, S_{min}, S_{mean} \rangle, & \text{if } P_{max} < P_{min} < P_{mean} \\ \langle S_{mean}, S_{max}, S_{min} \rangle, & \text{if } P_{mean} < P_{max} < P_{min} \\ \langle S_{mean}, S_{min}, S_{max} \rangle, & \text{otherwise} \end{cases} \quad (2.14)$$

Table 2.2 Comparison between pattern detections methods

Method	Advantages	Disadvantages
Matrix template pattern	Gives good results when identifying trends or trend inversions, because the noise removal associated with the method that smooth the data allows to clearly identify these market movements	With large datasets and low dimensional reduction can be less effective, since has to convert to matrixes the financial time series and them multiply two matrixes to take any decision. Used only on pattern recognition
Perceptually important points (PIP)	Even when applying high levels of dimensional reduction to the data is possible to preserve the important information in the financial time series. This method was created to deal with financial data	Hard to use as pattern discover method, is mainly used to pattern recognition. The data has to be converted to several representations corresponding to the same dimensions of the existing templates that the method is trying to identify
Symbolic aggregate approXimation (SAX)	Converts data to a symbolic sequence, allowing to easily identifying patterns by comparing strings. Easy to implement and fast converting the data to the symbolic representation	Tends to lose important information of the financial time series if the method conversion parameters are not well chosen

where,

S The SAX symbol;

P Position in the horizontal axes where the maximum, minimum and mean points occur.

In Fig. 2.34 is an example of this method representation, where the use of the extra symbols allows to preserve the information about the oscillations in each of the segments.

A problem detected in the SAX representation method when using financial time series is the loss of some important points, because smoothes the data since is based on a mean method. To overcome this problem it could be suggested the use of eSAX, but the additional symbols correspond to a lower dimensional reduction of the time series, this would mean more data to be handled. It is clear that the dimensional reduction that can be achieved and the discretization of data in amplitude by the basic SAX method are characteristics that could be used to create an efficient algorithm to pattern discover. To solve the problem of the smooth data it is possible to try to choose the method parameters in a more careful way, for instance considering smaller segments when converting the time series.

Table 2.3 Market forecasting methods and results

References	Years	Method	Used data	Financial market	Period	Algorithm performance
[15]	2008	Bull flag pattern w/matrix template	Stock price	NYSE composite. index	1967–2003	4.59 % (Transaction average over the period)
[31]	2007	Bull flag pattern w/matrix template	Stock price	NASDAQ	1985/04/ 03–2004/ 03/20	4.38 % (Transaction average over the period)
[16]	2002	Hybrid neural network w/ pattern detection	Stock price and vol.	NYSE composite Index	1984/07/ 24–1998/ 06/11	66 % (Days market goes up after buying order)
[32]	2006	Template-based	Stock price	Several	N/A	96 % (Hits on pattern identification)
[32]	2006	Rule-based	Stock price	Several	N/A	38 % (Hits on pattern identification)
[33]	2009	ENN + PLR	Stock price	Google.com	2007/05/ 01–2007/ 10/19	95 % (Average rate of return)
[34]	2003	AANN	Stock price	Kospi 200 futures	2001	31 % (year profit)
[35]	2009	ANN	Stock price	Stocks from Bovespa	2008/05/ 02–2008/ 12/02	Worst return 23 % Best return 130 %
[36]	2008	SOM GA-BPN	Several	BSE-30 index	2007/11/ 12–2008/ 01/01	30 % more return than the index
[37]	2008	GACS-M	Stock price	Stocks from TAIEX	2000/09/ 21–2007/ 09/21	21,42 % (Average rate return)
[18]	2010	GA	Several	Nikkei 225	Jan.1999– Dec.2009	57,4 % (Profit rate)
[38]	2010	GA-ANN	Stock price	Shenzhen	N/A	0,7176 (Correlation between prediction and actual value)

(continued)

Table 2.3 (continued)

References	Years	Method	Used data	Financial market	Period	Algorithm performance
[39]	2011	MCS RBFNN	Stock price	Hang Seng index	10 years	167 (Average earning)
[40]	2011	NN	Price	Forex	2010/09/20–2011/01/21	0.666186e-3–0.101144e-2 (Mean square error)
[20]	2011	HLP-ANN	Stock price	Shanghai index	1991/11/18–2009/02/10	7.16 % −9.65 % (Prediction error)
[41]	2010	WMM + Kalman filter	Daily trading volume	Bowin Technology & Denghai Seed industry	2008/02/13–2009/02/13	0.13–0.1879 (SNR—Prediction)
[42]	2010	GRA + GNP	Stock price	Several	2005/01/08–2007/01/04	4 % (Average monthly return rate)

2.2.2 Why Choosing GA and SAX

In Sect. 2.2.1.3 the GA was chosen as the solution with the ability to discover patterns. The easy way of coding new patterns on the chromosome genes and then evolve those patterns to match important patterns existent on the financial data, makes this optimization tool the ideal to complete this task with success. Now it is necessary a way to represent the time series and patterns efficiently in order to use this form of representation in the GA. In Table 2.2 is presented a comparison between the previous time series representation methods, PIP, Matrix and SAX.

From Table 2.2 and as was referred in Sect. 2.2.1.2, the matrix representation has large limitations, it is only capable of supporting simple graphic formations, so for the discover process is not an interesting choice and is left out. The PIP representation has the advantage of being born to represent financial data, but the problem of having to keep several dimensional representations of the same data accordingly to the complexity of the patterns, makes this method a more complex to adapt to the GA. Finally SAX, which uses a symbolic representation of the time series, proves to be ideal to be coded in the gene format. The symbolic discretization, which applies in the vertical axes of the data, creates a symbol that is possible to save on chromosome gene and easily manipulated in the crossover and mutation process. Also the discretization factor will help to reduce the overfitting, since the pattern terms will be chosen from a limited set of values rather than a real number interval. The problem identified in this representation, the loss of some important information in the financial time series, could be overcome by the inclusion of the adjustable SAX representation factors as parameters to be optimized by the genetic algorithm.

2.3 Conclusions

As was previously referred, chart pattern detection is an important method of trying to forecast financial markets. Due to the difficulty of this task several methods, using technical or fundamental indicators or even the price of financial data, were used to forecast markets, some of those methods results are referred on Table 2.3. The lack of solutions that tries to predict market behavior based in graphic pattern discovery is notorious, so it is reasonable to think that a solution on this area could bring a natural advantage on financial trading. The methods presented in Sect. 2.2.1, based on pattern recognition, did not offer the ability of discovering new pattern formations and in the present market conditions to be able to discover new patterns is an important characteristic, because allows the forecast method to adapt and learn new graphic pattern formations, which could be relevant to the trading decision support system. Even all the works about the SAX methodology used this method as a pattern recognition tool. In areas other than financial, SAX has been used to discover patterns, working with K-means or K-motifs as are called in [24].

In the discovery pattern area one of the goals is to find patterns with algorithms more efficient. One tool largely used to find solutions in an effective way is the GA. So, now what is needed is a data time series representation that could work together with the GA in an efficient way and could be represented in a chromosome format. From the several data representation studied the first choice was SAX, since the symbolic representation of data could be easily inserted on genes of a chromosome structure and assists the GA in finding new and important graphic pattern formations.

References

1. B. Baumohl, *The Secrets of Economic Indicators: Hidden Clues to Future Economic Trends and Investment Opportunities*, 2nd edn. (Wharton School Publishing, Pennsylvania, 2007)
2. S.B. Achelis, *Technical Analysis From A-To-Z.* (Vision Books, New Delhi, 2000)
3. T.N. Bulkowski, *Encyclopedia of Chart Patterns*, 2nd edn. (Wiley, New Jersey, 2005)
4. R. Fischer, J. Fischer, *Candlesticks, Fibonacci, and Chart Pattern Trading Tools: A Synergistic Strategy to Enhance Profits and Reduce Risk.* (Wiley, New Jersey, 2003)
5. R.W. Colby, *The Encyclopedia of Technical Market Indicators*, 2nd edn. (McGraw-Hill, New York, 2003)
6. E. Ponsi, *Forex Patterns & Probabilities—Trading Strategies for Trending and Range-Bound Markets.* (Wiley Trading, Wiley, New Jersey, 2007)
7. P. Parracho, R. Neves, N. Horta, Trading with optimized uptrend and downtrend pattern templates using a genetic algorithm kernel. IEEE Congr. Evol. Comput. 1895–1901 (2011). doi:10.1109/CEC.2011.5949846
8. T.-C. Fu, A review on time series data mining. Int. J Eng. Appl. Artif. Intell. **24**(3), 164–181 (2011). ISSN 0952-1976. doi:10.1016/j.engappai.2010.09.007
9. F.-L. Chung, T.-C. Fu, R. Luk, V. Ng, Flexible time series pattern matching based on perceptually important points. Int. Jt. Conf. Artif. Intell. Workshop on Learn from Temporal and Spatial Data, 1–7 (2001)
10. T.-C. Fu, F.-L. Chung, R. Luk, C.M. Ng, Representing financial time series based on data point importance. Eng. Appl. Artif. Intell. **21**(2):277–300 (2008). ISSN 0952-1976. doi:10.1016/j.engappai.2007.04.009
11. F.-L. Chung, T.-C Fu, V. Ng, R.W.P Luk, An evolutionary approach to pattern-based time series segmentation. IEEE Trans. Evol. Comput. **8**(5), 471–489 (2004). doi:10.1109/TEVC.2004.832863
12. T.-C. Fu, F.-L. Chung, R. Luk, V. Ng, Pattern discovery from stock time series using self-organizing maps. The 7th ACM SIGKDD International Conference on Knowledge Discovery and Data Mining (2001)
13. W. Leigh, , N. Modani, R. Purvis, T. Roberts, Stock market trading rule discovery using technical charting heuristics. Expert Sys. Appl. **23**(2), 155–159 (2002). ISSN 0957-4174. doi:10.1016/S0957-4174(02)00034-9
14. W. Leigh, N. Paz, R. Purvis, Market timing: a test of a charting heuristic. Econ. Lett. **77**(1), 55–63 (2002). ISSN 0165-1765. doi:10.1016/S0165-1765(02)00110-6
15. W. Leigh, C.J. Frohlich, S. Hornik, , R.L. Purvis, , T.L. Roberts, Trading with a stock chart heuristic. IEEE Trans. Part A: Sys. Hum. Sys. Man Cybern. **38**(1), 93–104 (2008). doi:10.1109/TSMCA.2007.909508

16. W. Leigh, M. Paz, R. Purvis, An analysis of a hybrid neural network and pattern recognition technique for predicting short-term increases in the NYSE composite index. Omega, Elsevier **30**, 69–76 (2002)

17. W. Leigh, R. Purvis, J.M. Ragusa, Forecasting the NYSE composite index with technical analysis, pattern recognizer, neural network, and genetic algorithm: a case study in romantic decision support. J Decis. Support Sys. **32**(4), 361–377 (2002). doi:10.1016/S0167-9236(01)00121-X

18. K. Matsui, H. Sato, Neighborhood evaluation in acquiring stock trading strategy using genetic algorithms. The International Conference on Soft Computing and Pattern Recognition (SoCPaR), pp. 369–372 (2010). doi:10.1109/SOCPAR.2010.5686733

19. J. Pinto, R. Neves, N. Horta, *Fitness function evaluation for MA trading strategies based on genetic algorithms*. ed. by N. Krasnogor. Proceedings of the 13th Annual Conference Companion on Genetic and Evolutionary Computation (GECCO '11). ACM, New York, NY, USA, pp. 819–820 (2011). doi:10.1145/2001858.2002140

20. L. Wang, Q. Wang, Stock market prediction using artificial neural networks based on HLP. Int. Conf. Intel. Hum. Mach. Sys. Cybern. **1**, 116–119 (2011). doi:10.1109/IHMSC.2011.34

21. L.J. Cao, F.E.H. Tay, Support vector machine with adaptive parameters in financial time series forecasting. IEEE Trans. Neural Netw. **14**(6), 1506–1518 (2003). doi:10.1109/TNN.2003.820556

22. D. Zhang, H. Song, P. Chen, Stock market forecasting model based on a hybrid ARMA and support vector machines. Proceedings of the 15th International Conference on Management Science and Engineering, Long Beach, USA (2008)

23. J. Lin, E. Keogh, S. Lonardi, B. Chiu, A symbolic representation of time series, with implications for streaming algorithms. Proceedings of the 8th ACM SIGMOD International Conference on Management of Data, Workshop on Res. Issues in Data Mining and Knowledge Discovery, pp. 2–11 (2003). doi:10.1145/882082.882086

24. J. Lin, E. Keogh, S. Lonardi, P. Patel, Finding motifs in time series. Proceedings of the 8th ACM SIGKDD International Conference on Knowledge Discovery and Data Mining . 2nd Workshop on Temporal Data Mining, pp. 53–68 (2002)

25. C. Faloutsos, M. Ranganathan, Y. Manolopoulos, Fast subsequence matching in time-series databases. Proceedings of the ACM SIGMOD International Conference on Management of Data, Minneapolis, pp. 419–429 (1994). doi:10.1145/191839.191925

26. E. Keogh, K. Chakrabarti, M. Pazzani, S. Mehrotra, Dimensionality reduction for fast similarity search in large time series databases. J Knowl. nf. Sys. (2000). doi:10.1145/191839.191925

27. J. Lin, E. Keogh, L. Wei, S. Lonardi, Experiencing SAX: A novel symbolic representation of time series. Data Min. Knowl. Discov. **15**(2), 107–144 (2007). doi:10.1007/s10618-007-0064-z

28. J. Shieh, K. Keogh, iSAX: Indexing and mining terabyte sized time series. Proceedings of the 14th ACM SIGKDD International Conference on Knowledge Discovery and Data Mining, New York, NY, USA, pp. 623–631 (2008). doi:10.1145/1401890.1401966

29. A. Camerra, T. Palpanas, J. Shieh, E. Keogh, iSAX 2.0: Indexing and mining one billion time series. Proceedings of the 2010 IEEE International Conference Data Mining (ICDM '10). IEEE Computer Society, Washington, DC, USA, pp. 58–67 (2010). doi:10.1109/ICDM.2010.124

30. B. Lkhagva, Y. Suzuki, K. Kawagoe, New time series data representation ESAX for financial applications. Proceedings of the 22nd International Conference Data Engineering Workshops (ICDEW '06), p. 115 (2006). doi:10.1109/ICDEW.2006.99

31. J.-L. Wang, S.-H. Chan, Stock market trading rule discovery using pattern recognition and technical analysis. Expert Sys. Appl. **33**(2), 304–315 (2007). doi:10.1016/j.eswa.2006.05.002

32. T.-C. Fu, F.-L. Chung, R. Luk, C.M. Ng, Stock time series pattern matching: Template-based vs rule-based approaches. Eng. Appl. Artif. Intell. **20**(3), 347–364 (2007)

33. P.-C. Chang, C.-Y. Fan, C.-H. Liu, Y.-W. Wang, J.-J. Lin, Evolving neural network with dynamic time warping and piecewise linear representation system for stock trading decision

making. 2009 WRI World Congr. Comput. Sci. Inf. Eng. **5**, 303–307 (2009). doi:10.1109/CSIE.2009.36

34. L. Junmyung, C. Sungzoon, B. Jinwoo, Trend detection using auto-associative neural networks: Intraday KOSPI 200 futures. Proceedings of the IEEE International conference on Computational Intelligence for Financial Engineering, pp. 417–420. (2003) doi:10.1109/CIFER.2003.1196290

35. L.C. Martinez, D.N. da Hora, J.R. de M Palotti, W. Meira, G.L. Pappa (2009) From an artificial neural network to a stock market day-trading system: A case study on the BM&F BOVESPA, The International Joint Conference on Neural Networks 2006–2013 (2009). doi:10.1109/IJCNN.2009.5179050

36. A.U. Khan, T.K. Bandopadhyaya, S. Sharma, Classification and identification of stocks using SOM and genetic algorithm based backpropagation neural network. International Conference on Innovation in Management and Information Technology, pp. 292–296 (2008). doi:10.1109/INNOVATIONS.2008.4781644

37. P.-C. Ko, P.-C. Lin, C.-S. Shih, Stock valuation and dynamic asset allocation with genetic algorithm and cubic spline. Int. Conf. Mach. Learn. Cybern. **7**, 3997–4000 (2008). doi:10.1109/ICMLC.2008.4621101

38. H.-N. Hao, Short-term forecasting of stock price based on genetic-neural network. 6th Int. Conf. Nat. Comput. **4**, 1838–1841 (2010). doi:10.1109/ICNC.2010.5584528

39. W.W.Y. Ng, X.-L. Liang, P.P.K. Chan, D.S. Yeung, Stock investment decision support for Hong Kong market using RBFNN based candlestick models. Int. Conf. on Mach. Learn. Cybern. **2**, 538–543 (2011)

40. H. Tahersima, M. Tahersima, M. Fesharaki, N. Hamedi, Forecasting stock exchange movements using neural networks: A case study. Proceedings of the International Conference on Future Computer Science and Applications, pp. 123–126 (2011)

41. Z. Fang, G. Luo, F. Fei, S. Li, Stock forecast method based on wavelet modulus maxima and kalman filter. Proceedings of the 4th International Conference on Management of e- Commerce and e-Government, pp. 50–53 (2010). doi:10.1109/ICMeCG.2010.19

42. V. Parque, S. Mabu, K. Hirasawa, Enhancing global portfolio optimization using genetic network programming. Proceedings of the SICE Annual Conference, pp. 3078–3083 (2010)

Chapter 3
SAX-GA Approach

Abstract In this chapter an innovative methodology for pattern discovery in financial time series will be presented. The combination of SAX representation method with the use of GA to optimize the search and creation of new investment strategies will be explored. Taking advantage of the symbolic representation and dimensional reduction provided by SAX, several financial time series will be analyzed in order to search for meaningful patterns to reveal periods of time to invest on the stock market. The search and investment criteria will be defined and optimized by the use of GA, several chromosomes structures were considered in order to provide more accurate results. Basically two approaches will be presented, a first one will try to discover patterns that indicate a bull market condition, in order to invest long; another approach will combine the previous one with the detection of patterns signaling a bear market, to invest short and long. These two approaches will be investigated and compared in the results section.

Keywords Symbolic Aggregate Approximation (SAX) · Piecewise Aggregate Approximation (PAA) · Genetic Algorithm (GA) · SAX-GA Method

3.1 SAX Representation for Time Series

Like was described in Chap. 2, SAX stands for Symbolic Aggregate approximation, which is an evolution of Piecewise Aggregate approximation (PAA), where the time series are converted to a symbolic representation of data instead of a numeric one like in PAA. So, as an evolution of PAA the first steps of the method are the same, where basically a time series is divided in equal parts and each of the parts becomes represented by the arithmetic mean of the points that lay in the respective part.

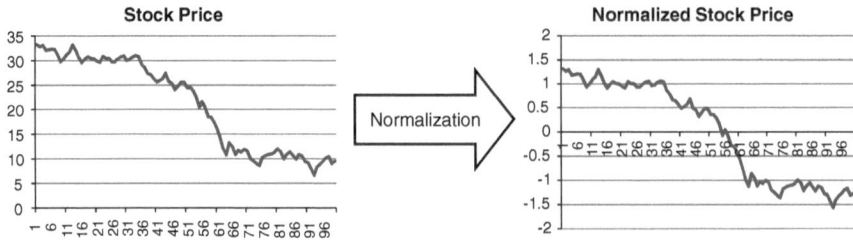

Fig. 3.1 Normalization process of a stock quote time series

In order to discover patterns in large time series, it is necessary to break the data into smaller time series windows; this will allow an easier comparison between windows in order to evaluate the similitude degree among data windows. In the next sections this method will be presented in detail.

3.1.1 Normalization and PAA Method

Financial time series are usually rather large, it is normal to analyze data where the price is sampled every second, so to create a process to be able to deal with this large datasets more effectively is common to divide the dataset into smaller ones. So, to discover patterns in large time series of dimension m this dataset must be break into smaller time series windows of size $n \ll m$. These windows must be compared with each other, so the characteristics of these smaller time series must be similar, same magnitudes and base line. Therefore to apply this transformation to the windows, data has to be normalized, Eq. 3.1. This normalization does not affect the original shape [1], and scales the data to the same relative magnitude Fig. 3.1.

$$x'_i = \frac{x_i - \mu_x}{\sigma_x} \qquad (3.1)$$

where,

x_i Points in window W_k;

μ_x is the mean of the points in W_k;

σ_x is the standard deviation of all the x_i points.

After normalization the data windows are ready to be compared, but the dimension of this data is high. At this point no data has been removed from the original time series, turning this process very expensive in time and computational resources. So some method of dimensionality reduction is needed, as it was said before SAX is based on PAA to achieve this objective.

In PAA the time series windows are divided in w equal size segments and each segment is represented by the arithmetic mean of the points in it, according to Eq. 3.2.

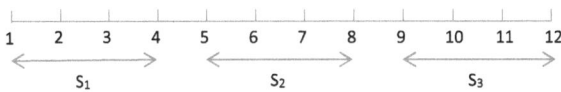

Fig. 3.2 Size 12 window divided in 3 segments, each point contributes to one segment only

Fig. 3.3 Size 12 window divided in 5 segments, the points between segments contribute to the neighbors segments

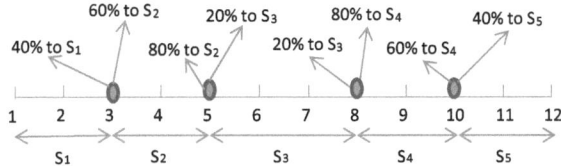

$$\bar{x}_j = \frac{w}{n} \sum_{i=\frac{n}{w}(j-1)+1}^{\frac{n}{w}j} x_i' \tag{3.2}$$

where,

w number of segments that divide the window;

n size of the window;

x_i' points in the window.

The Eq. 3.2 is valid if the relation n/w has an integer result, in this case each point contributes entirely to the frame or segment where is inserted, which is the case of the example in Fig. 3.2.

In the case of a non integer relation, the point in the frontier between segments must contribute with some part to each of the segments, this method was developed by Li Wey,[1] as shown in Fig. 3.3.

From the example in Fig. 3.3, where there are 12 points and 5 segments, it means that each segment must have the contribution of 2.4 points for the mean. So, points 1 and 2 entirely belong to segment S1 and to complete the S1 segment 40 % of point 3 must be added. For calculating the S2 segment, the remain 60 of point 3 must be added, the entire point 4 and 80 % of point 5 to complete the 2.4 points per segment. The rest of the segments are completed according to this logic.

These two methods of calculation, accordingly to the compression factor n/w, allow representing the time series by a smaller of numbers corresponding to the arithmetic mean of the points in a segment. Basically this describes the PAA dimensional reduction method, it is possible to verify in Fig. 3.4 that each segment is represented by the mean of points in it.

---

[1] http://alumni.cs.ucr.edu/~wli/

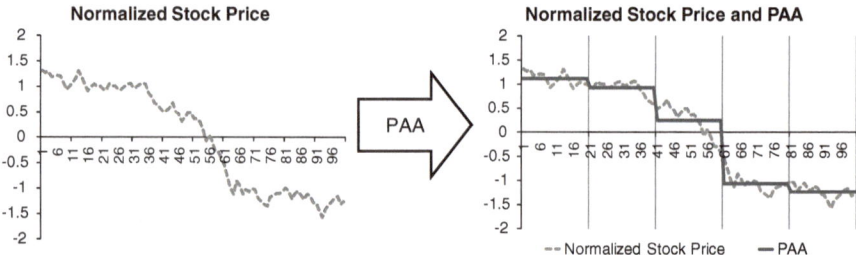

Fig. 3.4 PAA dimensional reduction method

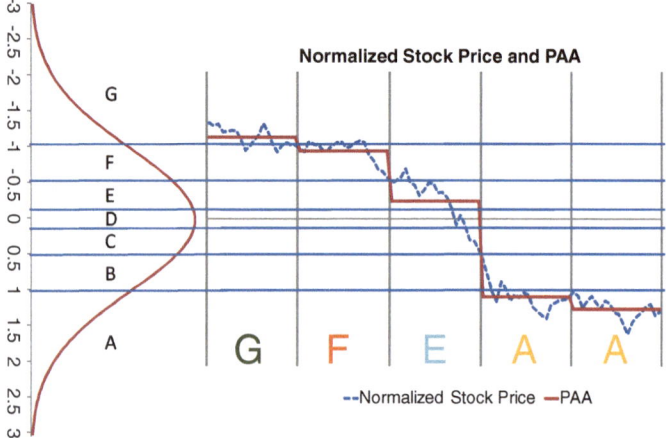

Fig. 3.5 SAX representation

3.1.2 Symbolic Representation

After getting the PAA transformation, to convert the time series to a symbolic representation, the amplitude of the time series must be divided into a intervals, and to each of them is assigned a symbol. In order to produce equiprobable intervals and since the data has been normalized, a normal distribution curve will be applied to the vertical axis and breakpoints are calculated to produce equal areas under the curve, Fig. 3.5.

The use of this curve allows to get a more detail quantization levels around the mean, where is more probable to have the segments. In [2] is presented that although some time series that do not have a normal statistical distribution, when a smaller data subset, like the windows presented here, are extracted, these data windows reveal to have a normal distribution or are very close to this distribution.

---

² http://www.cs.ucr.edu/∼eamonn/SAX.htm

Table 3.1 Breakpoints vs. symbol size

	a = 2	a = 3	a = 4	a = 5	a = 6	a = 7	a = 8
β_1	0	−0.4307	−0.6745	−0.8416	−0.9674	−1.0676	−1.1503
β_2	−	0.4307	0	−0.2533	−0.4307	−0.5659	−0.6745
β_3	−	−	0.6745	0.2533	0	−0.18	−0.3186
β_4	−	−	−	0.8416	0.4307	0.18	0
β_5	−	−	−	−	0.9674	0.5659	0.3186
β_6	−	−	−	−	−	1.0676	0.6745
β_7	−	−	−	−	−	−	1.1503

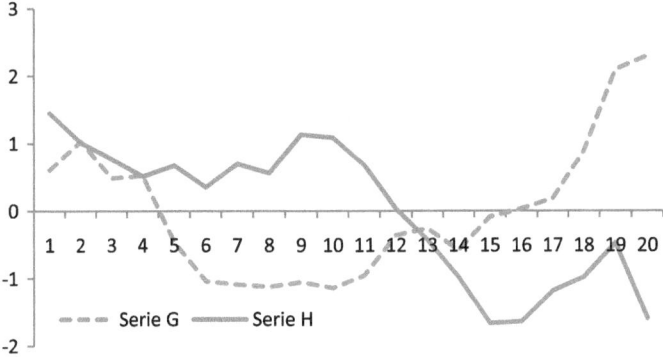

Fig. 3.6 Two time series comparison

After applying the curve and determine the breakpoints, each segment is evaluated to determinate to which interval belongs. The PAA level defined which symbol is assigned to represent that segment. Applying this method to all the segments and all the windows will produce sequences of symbols, which now represents the time series.

The β breakpoints can be obtained from statistical books or like in the case of Table 3.1 from the Matlab© code available in the SAX official web site.[2] In Table 3.1, are the breakpoints values according to the a number of symbols or alphabet size, used in time series representation.

3.1.3 Pattern Identification: Distance Measure

As said before, several windows must be compared in order to identify patterns, as result of those comparisons a distance value between two time series is returned, the lower distance between windows reveal a greater degree of similitude, so a pattern is probable present. In Fig. 3.6, two normalized windows time series are being compared, the regular way to calculate de distance between them is by the Euclidian distance between points, Eq. 3.3.

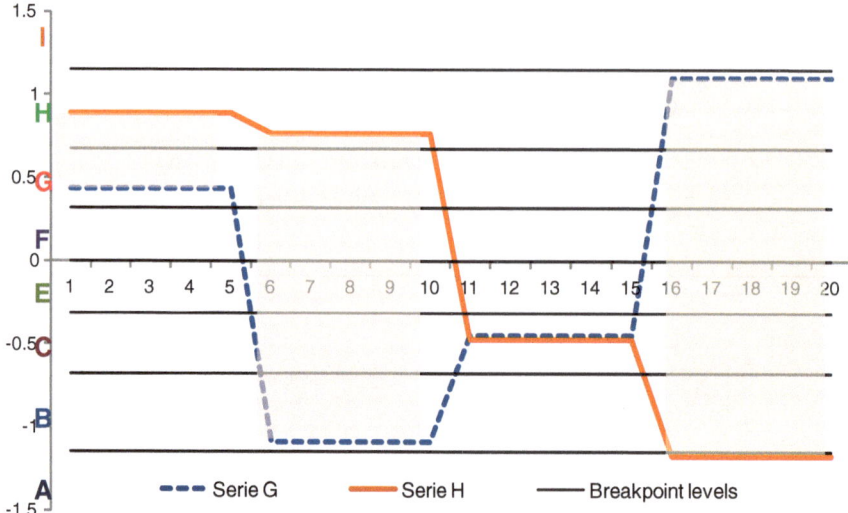

Fig. 3.7 Distance between two time series in PAA representation

Fig. 3.8 Distance
calculation between two SAX
sequences

$$D_{EUC} = \sqrt{\sum_{i=1}^{n} (G_i - H_i)^2} \qquad (3.3)$$

where,
G_i is the i point of the time series G;
H_i is the i point of the time series H;
n number of points in the window.

If the two windows time series are converted to PAA, Fig. 3.7, is also possible to calculate a distance to make a comparison, in this case is possible to obtain a lower bounding distance based on the Euclidian distance given by Eq. 3.4.

$$D_{EUC}^{PAA} = \sqrt{\frac{n}{w}} \sqrt{\sum_{i=1}^{w} (G_i^{PAA} - H_i^{PAA})^2} \qquad (3.4)$$

where,
G_i is the i point of the time series G^{PAA};
H_i is the i point of the time series H^{PAA};
n number of points in the window;
w number of segments.

In case the PAA representations are converted to symbolic SAX sequences, Fig. 3.8, using the alphabet symbols according to the representation on the right side of Fig. 3.7, is also possible calculate a lower bounding distance defined by Eq. 3.5.

$$MINDIST(G^{SAX}, H^{SAX}) = \sqrt{\frac{n}{w}}\sqrt{\sum_{i=1}^{w}(dist(G_i^{SAX}, H_i^{SAX}))^2} \qquad (3.5)$$

where,
G_i is the i point of the time series G^{SAX};
H_i is the i point of the time series H^{SAX};
n number of points in the window;
w number of segments;
$dist(.)$ function given by Eq. 3.6.

$$dist(p_i, q_j) = \begin{cases} 0 \,|i-j| \leq 1 \\ \beta_{j-1} - \beta_i \, i < j-1 \\ \beta_{i-1} - \beta_j \, i > j+1 \end{cases} \qquad (3.6)$$

This last distance measure will be used to detect pattern in time series when using SAX, from Eq. 3.6 is clear that the parameter that affects the measure precision are the number of breakpoints that divide the window according the vertical axes, this division correspond to the alphabet size chosen to represent data. Several studies tried to identified the best value for this parameter [3], most of these studies reveal that this parameter is data dependable, in another study [4], it was found that an alphabet size of 5–8 symbols are values that could provide an higher efficiency from the computational point of view.

Equation 3.5 defines the usual distance measure used in the SAX methodology. The application of the GA, in order search for the minimum distance to positively identify a pattern, will allow to define other distance measures. One of these new measures is calculated by Eq. 3.7.

$$dist = \sqrt{\sum_{i=1}^{w}(T_i - P_i)^2} \qquad (3.7)$$

where,
w word size;
T_i symbol i of the time series;
P_i symbol i of the pattern.

Basically, this new distance measure, will calculate how close two SAX sequences are to each other. This method is faster than the standard MINDIST from Eq. 3.5, since the operation does not need to check the breakpoints table to do the calculations and is expected that the GA will adapt to this type of measure. An example of this measure is in Fig. 3.9, and takes advantage of the possibility that C++ can subtract *char* data type.

Fig. 3.9 Example of a
distance calculation, based on
the discrete symbolic
representation

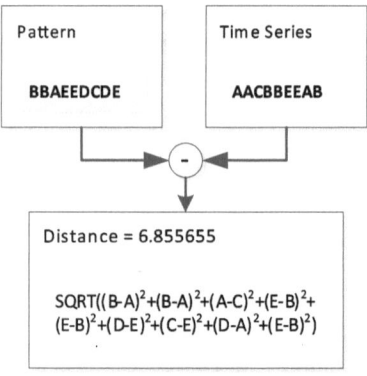

3.1.4 SAX Parameters Estimation

The SAX parameters used to define the representation of data and that could affect the correct pattern discovery are:

A—Number of intervals that divides the normal curve, which corresponds to the number of symbols in the alphabet;

N—Window size, that corresponds to the number of data points that will be converted to a SAX sequence;

W—Word size, which is the number of symbols in a sequence that represents a window time series.

In order to estimate the best values for these parameters, it was selected a financial time series from an S&P500 stock, with almost 3,100 points in the period of 1998 to 2010. With this data was made an exhaustive search of patterns, with several combinations of values. This exhaustive pattern search process, take values, for the several parameters, from the following intervals:

$n \in [10 \ldots 150]$

$w \in [2 \ldots n/2]$

$a \in [2 \ldots 20]$

For the parameter estimation an application on C++, that converts the time series into SAX sequences using the combinations of values defined previously, was developed [5]. For each combination, the number of different patterns detected and the number of occurrences of those patterns in the time series was evaluated. In this stage, the patterns are sequences of symbols exactly equal, by the definition of distance given in the previous Sect. 3.1.3, the distance between sequences could be zero even if they are not exactly the same. But for this test, only patterns with the same sequence of symbols were considered. This will simplify and optimize the application, since the patterns are saved as the *key* on a *map* associative container, this makes it easy to identify patterns without having to calculate distances between them. After running the several combinations, the data was analyzed in Matlab©.

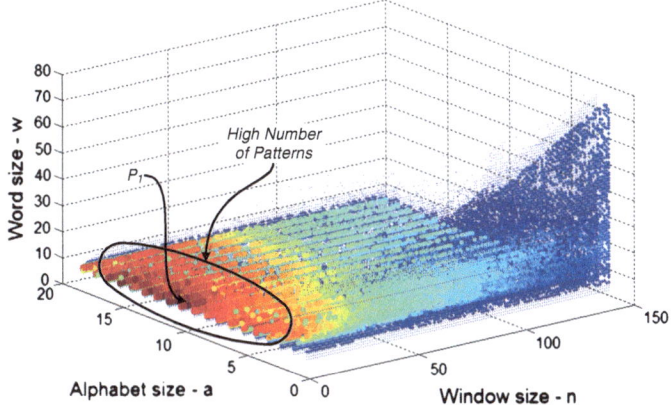

Fig. 3.10 Different patterns surface map

• Different patterns discovery

In Fig. 3.10 is a representation of the number of different patterns found, it is possible to identify (inside the ellipse/red color) areas where the parameters tested reveal a larger number of patterns. The ellipse/red color, indicates a higher number of patterns present. It is possible to verify, by the scatter points/dark blue, that patterns with bigger window size and word size only exists with small alphabet size, this makes sense since complex and longer patterns should be more difficult to find.

The maximum number of different patterns identified in Fig. 3.10 was 597, the point where the value is obtained:

$$p_1 = \begin{cases} n = 17 \\ w = 4 \\ a = 13 \end{cases}$$

This point will be tested at the search for patterns with the GA, it is reasonable to think that in an area with a large number of different patterns some of them will be important solutions to the problem in analysis.

• Patterns Occurrences

Another analysis made to the data is the total number of occurrences of the patterns found for each combination of parameters, in Fig. 3.11 a surface representing this fact is shown. For this analysis the maximum value was 2898, at the point:

$$p_2 = \begin{cases} n = 10 \\ w = 3 \\ a = 18 \end{cases}$$

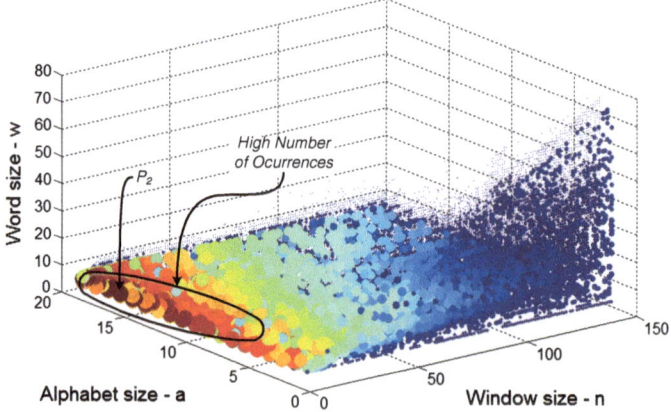

Fig. 3.11 Pattern occurrences surface map

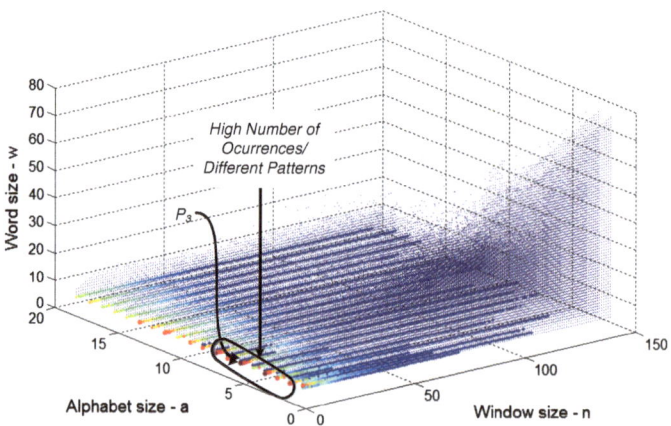

Fig. 3.12 Relation between patterns found and occurrences

The point p_2 needs to be explored, since areas where a large number of patterns occur is probable that relevant solutions appear.

- Patterns importance level

A third analysis was made, this one combines the first two, here is studied the "importance" of the patterns, for some combination of parameters. If a low number of patterns appear for a large number of occurrences is probably more significant than having many patterns with a small number of occurrences, the surface for this analysis is on Fig. 3.12.

In this analysis was identified the value 239.4, for the next parameters combination values:

Fig. 3.13 Simple chromosome structure used by the GA

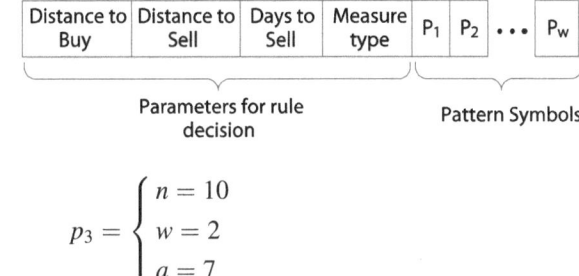

Distance to Buy	Distance to Sell	Days to Sell	Measure type	P₁	P₂	⋯	Pw

$$\underbrace{\hspace{5cm}}_{\substack{\text{Parameters for rule} \\ \text{decision}}} \qquad \underbrace{\hspace{3cm}}_{\text{Pattern Symbols}}$$

$$p_3 = \begin{cases} n = 10 \\ w = 2 \\ a = 7 \end{cases}$$

This last point, p_3 does not appear to be very relevant as patterns parameters, because patterns with a two letter word probably will not be very important.

Several more time series were subject to the same exhaustive search, and the results were quite similar, the same areas were identified. From the values of this three points is possible to conclude that areas with small words and large alphabets, are probably important for pattern discovery. This does not exclude the test and exploration of other areas in the solution space, because this test only identifies areas of high intensity and it may exists important patterns in areas of fewer density patterns.

3.2 SAX-GA Simple and Extended Chromosome Structure

This approach tries to identify uptrend patterns and application rules for those patterns. As the algorithm analyzes the financial time series, with the use of a sliding window of size defined by the n SAX parameter, must generate buying orders when the pattern is present and sell orders when the pattern disappears. In order to reduce the risk, another factor was introduced in the transaction process, this parameter will define for how many days the algorithm stays in the market. So, when the number of days in the buying position exceeds some threshold, the algorithm will close the position.

Based on this description, the genetic algorithm should produce patterns and detect if they are present on the time series. Since the SAX representation is used, the patterns are sequences of symbols and the distance from the pattern to the time series must be calculated to identify their presence. This fact brings the need to find how close the time series should be to the pattern, in order to justify the buying decision, as well the algorithm must identify how far should be to issues a sell order. The GA will use two distance measures, the first is the *"MINDIST"* from Eq. 3.5 presented in Sect. 3.1.3, the second measure, Eq. 3.7, takes advantage of the discrete symbolic representation and how far the symbols are from each other, and is expected to avoid some of the trivial matches identified in [3], since the distance between neighbors symbols will be different from zero.

Based on the previous definition, of how the GA should behave, the chromosome presented in Fig. 3.13 is the structure used by the population.

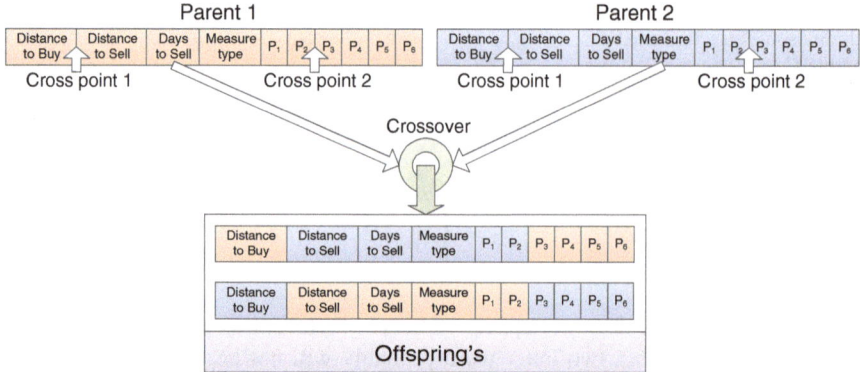

Fig. 3.14 Crossover process

The chromosome is divided in two major parts, the first one are the parameters that support the decisions of buy and sell, here are the two distances from the time series to the pattern, that permits to evaluate if the pattern is present in order to buy (Distance to Buy) or if the pattern is no longer present and is time to sell (Distance to Sell), another gene defines after how many days should the algorithm sell if it is in buying position (Days to Sell). The final gene of this part, is a bit that identifies which of the measures expressions, should be used to evaluate the distances (Measure type). The second part of the chromosome, are the symbols that constitute the pattern sequence ($P_{1...w}$).

The selection process used is a "uniform ranking selection" [6], applied to the best half of the population and then uses a two point crossover to generate the offspring's, Fig. 3.14. The option for two points instead of a single point was made because of the structure of the chromosome, where the first point cuts the chromosome in the section of the rule parameters and the second in the pattern symbols area. A multiple point crossover, with more than 2 points, was also studied but the chromosome length is small, the term that could increase this measure is the word size, but as has been seen in Sect. 3.1.4 this parameter has values around 3, so the chromosome has a total length of 8 genes. The generation of new population will be elitist, the best chromosomes will be preserved. The mutation rate is of 10 %, in tests made in another study on pattern match in financial markets [7] proved good results with this value. In the present study tests made with a mutation rate of 2 and 5 % reveal lower results for the same generation's number.

The fitness function that the GA will optimize is the total earnings produced by the investment strategy, defined by the pattern and application rule associated with it. This optimization is possible thanks to the creation of new patterns by the GA and the fine tune of the parameters that define the investment rules

The application has two main parts, the first one a training period where the GA will optimize the parameters, in the second the application will receive the best chromosome from train and applied it for a test unknown period, to evaluate the

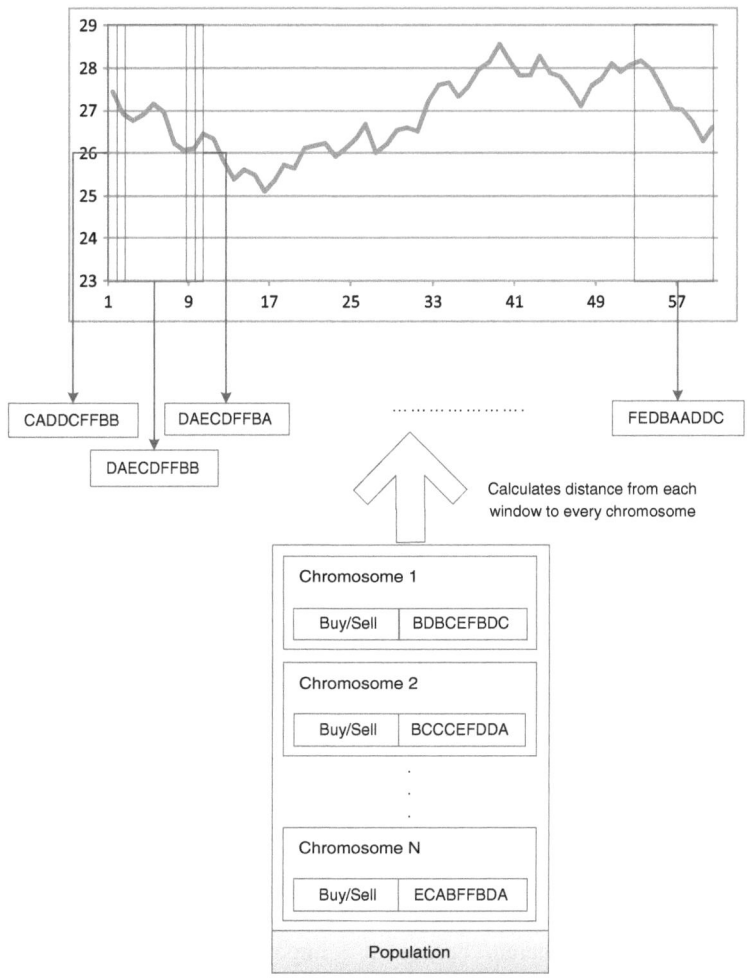

Fig. 3.15 Application process

result. For the simple chromosome structure, the SAX parameters are fixed and defined by the user, based on the study presented on Sect. 3.1.4.

From the functional point of view the program will slide a window along the time series and converts it to a SAX sequence. The patterns in the chromosomes will then be compared with each window sequence to calculate the distance and apply the rules defined in the chromosome, the distance to buy or sell, Fig. 3.15. If the distance is less than the *'Distance to buy'* defined in the chromosome the application will buy the stock. In case the stock has already been bought, the application sells the stock when the distance becomes large than the *'Distance to Sell'* gene or if the stock has been bought more days ago than the specified by the *'Days to Sell'* gene. At the end of the time series a new epoch begins and the

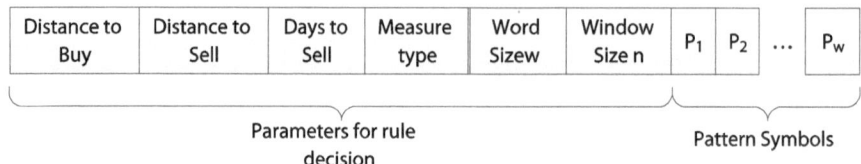

Fig. 3.16 Extended chromosome structure

process restarts with a new population that includes the best individuals and the new offspring's. The stop criteria used, is the end of improvement in the fitness function for several generations.

The basic difference from the training period to the testing period is that at the end of the of the time series no new generation is produced and the results are written for evaluation.

In the simple chromosome structure, the SAX parameters are defined at start based on the study presented before and in some experiments. In order to search for optimal values for the parameters and since the space search is rather large, it was created an extended chromosome structure, where two of these values will be optimized by the GA. The first is the number of symbols that define the pattern or word size w, the other parameter is the window size n. This new strategy causes a change in the chromosome structure, now represented by Fig. 3.16.

The additional genes did not change the crossover strategy, neither the mutation as the chromosome keeps the same basic structure than before, where still exist two major parts, the numeric parameters section and the symbolic pattern. In this new approach, and since now the chromosomes may have different length, the crossover points are different in each of the parents on the pattern section, Fig. 3.17.

The major change in this new structure relates to a constraint that must be added to the crossover process. This constraint guarantees that the SAX parameters in the offspring's are coherent, because it is impossible to have a SAX representation where the numbers of symbols, w, are greater than the data points in the window, n. This limitation will be presented on more detail in the next Sect. 3.3.

3.3 SAX-GA Multi-Chromosome Structure

Based with the potential of the SAX-GA method to find patterns, another approach has been made. Instead of just finding a pattern that chooses the moment to invest, this new strategy will find a pattern to enter long in the market, a pattern to leave the long position, a pattern to enter short and a pattern to exit the short position. So basically the previous strategy was extended and now the individuals are more evolved entities that are composed by four chromosomes, Fig. 3.18. The evolution to a multi-chromosome entity brings new problems and the possibility of

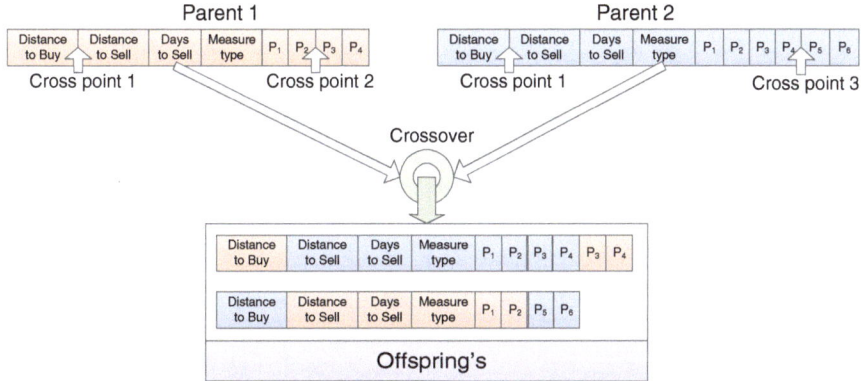

Fig. 3.17 Extended chromosome structure crossover

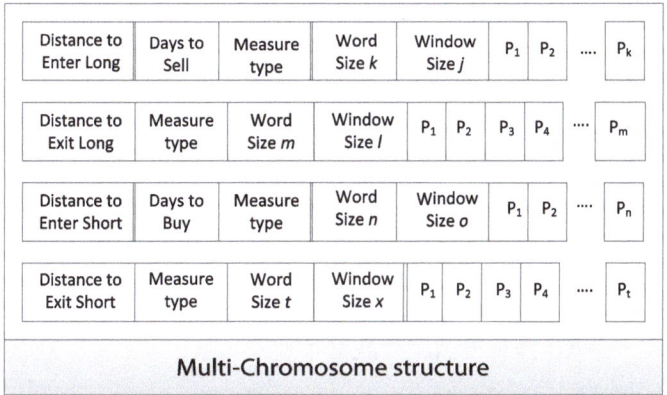

Fig. 3.18 Multi-chromosome structure

simultaneous contradictory signals of buy and sell that must be handled, since the algorithm now has two different strategies to invest.

As shown by Fig. 3.18 all the chromosomes may have different lengths and the number of parameters to optimize is now higher than in the previous strategies.

The major change made to the previous application was the crossover process, now the individuals have to exchange genetic material between homologous chromosomes. In this process each chromosome has different cross points since they have different number of genes, but the strategy of double point crossover is kept identical to the previous one. In the numeric parameters the cross points are identical in the homologous chromosomes, in the symbolic patterns the cross points are all different, in Fig. 3.19 a crossover example is shown.

From the example of Fig. 3.19 is expected that after each crossover the parameter *Word Size* must be evaluated since it depends of the new pattern produced, the number of symbols in the pattern is the word size, in fact this gene

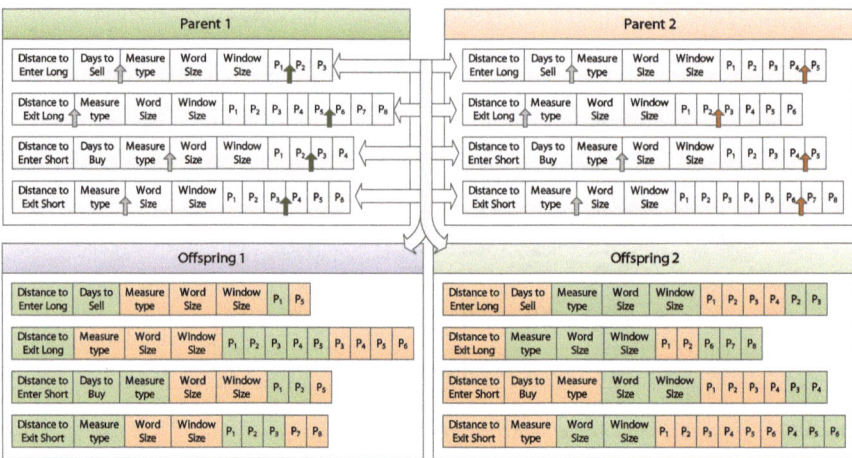

Fig. 3.19 Multi-Chromosome crossover example

could be considered redundant but it will be used in the next step of the crossover. After this first evaluation the *Window Size* is equally examined, because the *Word Size* cannot be higher than the *Window Size*, at least each point in the window corresponds to symbol of the word, so if *Word Size > Window Size* the window parameter must also be processed. The new *Window Size* is calculated using the previous relation between the *Window Size* and *Word Size* applied to the new *Word Size*, Eq. 3.8.

$$1^{st} Step \rightarrow C = \frac{Window\ Size_{Parent}}{Word\ Size_{Parent}}$$

$$2^{nd} Step \rightarrow Window\ Size_{Offspring} = C * Word\ Size_{Offspring} \tag{3.8}$$

This constrains are only applied if the values obtain in the crossover are inconsistent.

The mutation process also suffer some changes, the process used begins by randomly selects an individual, then for each chromosome randomly selects a gene from the parameter section and other from the pattern section to mutate its value. The mutation process in Fig. 3.20, shows that in case the selected gene is the *Word Size* parameter this mutation is discard, since is a parameter that depends on the number of symbols in the pattern section. In order to guarantee that the problem detected in the crossover, *Word Size > Window Size*, does not occur when the *Window Size* gene mutates, the mutation range values for this gene starts at *Word Size* value.

Another problem in this multi-chromosome methodology is related to the fact that now the algorithm can take two different decisions to invest, it could decide to enter long, buying the stock, or enter short selling it. To handle the case when the method generates the two signals for the same instant, the algorithm will decide which signal to apply based on the distance value from the time series to each

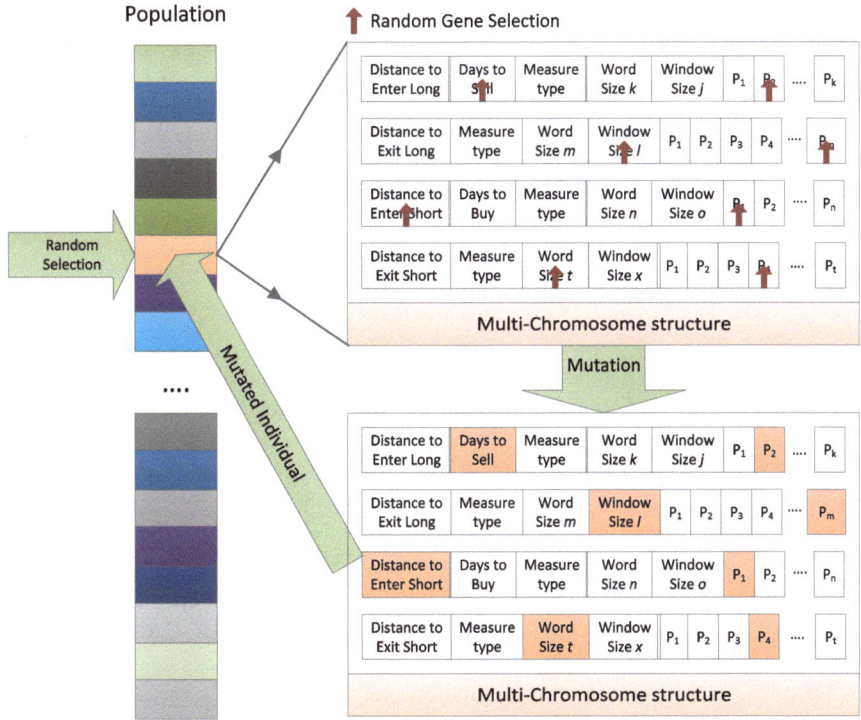

Fig. 3.20 Mutation process in the multi-chromosome population

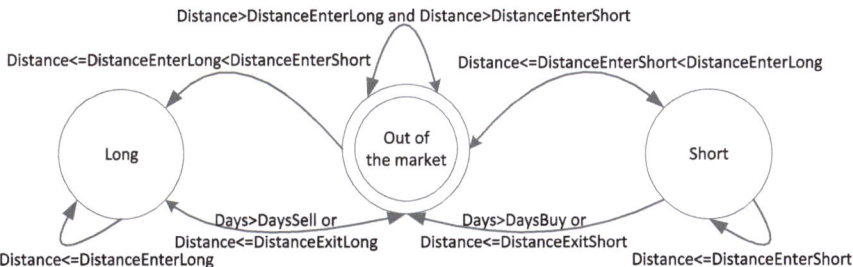

Fig. 3.21 Finite-state machine of the investment algorithm

pattern chromosome, enter long and enter short, and naturally the lower distance wins the tie.

After the initial buying or selling decision it is necessary to define when to leave the current position, to take this decision the algorithm will search for the pattern defined in the exit chromosomes, when it finds the new pattern sells or buys according to the previous state condition. In Fig. 3.21 a finite-state machine (FSM) that rules the different stages in the algorithm is presented. In this image the

Fig. 3.22 'ICMD' Pattern
used in the investment
strategy

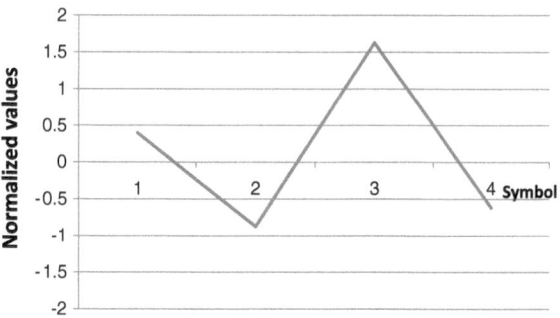

parameters Day and Distance are calculated at each slide of the window, the other
parameters are the values from the genes of the chromosome being examined.

3.4 SAX-GA in Action

In order to show how the algorithm implements the pattern detection and
investment rules, this section presents an example of the SAX-GA simple chromosome structure, Fig. 3.13, at work.

Next, the values used in this simulation are shown:

- Alphabet size: 13
- Window size: 17
- Word size: 4
- Distance to buy: 2.42876
- Distance to sell: 7.576
- Days to sell: 10
- Pattern to search: ICMD

In this test and due to the values chosen by the GA, the algorithm will close any
position when the number of days reach the *Days to sell* value, since the *Distance
to sell* value will never be reached, in fact this strategy always leave any investment position after the 10 days, for this specific simulation.

In Fig. 3.22, the pattern searched by the algorithm is presented.

The financial data used in this example is the S&P 500 index, and the period
examined is from January, 31 to April, 14 in the year 2005. This period in shown
in Fig. 3.23.

In Fig. 3.23 is possible to verify that two windows are defined, in the first one
the algorithm decides to invest and in the second stays out of the market. To better
understand the decision making by the algorithm, both windows will be examined
in detail next. Also, this Fig 3.23, shows the ROI in the dot/green line related to
the right graphic axes.

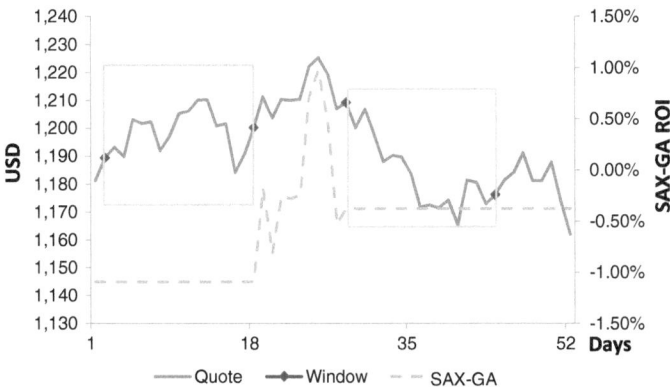

Fig. 3.23 S&P500 stock quote and SAX-GA investment return

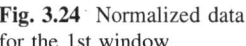
Fig. 3.24 Normalized data
for the 1st window

Fig. 3.25 Normalized data
for the 2nd window

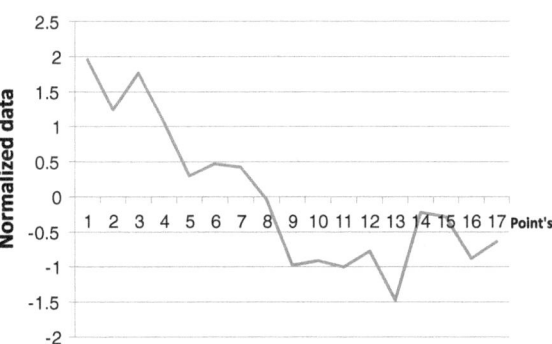

Like was previously presented, the first step of the SAX method is to normalize the data, in Figs. 3.24 and 3.25 the normalize data for the first and second window respectively, is shown.

After the normalization step, the data should be converted to a symbolic representation according to the SAX parameters values. This new representation will be now evaluated, in order to detect the degree of similitude between data and

Fig. 3.26 Distance between
pattern searched and data
sequence—1st window

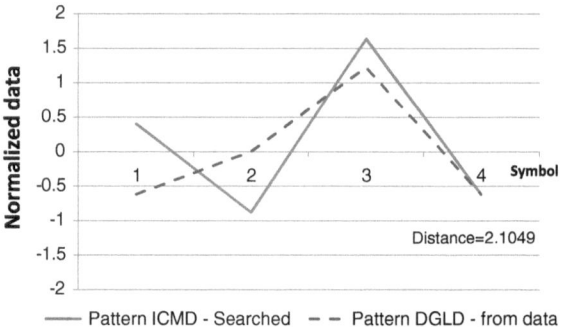

Fig. 3.27 Distance between
pattern and data sequence—
2nd window

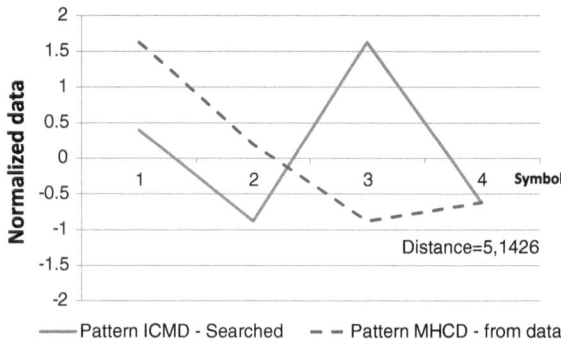

pattern searched, for this, the distance along the sequence is calculated. In
Fig. 3.26 the representation and distance to the pattern defined by the algorithm
are shown.

Since the distance calculated, for the first window, is below the *Distance to buy*
defined by the GA, the algorithm decides to enter the market by buying the asset.
As was stated before, in this investment strategy, the algorithm will leave the
market by selling the asset after 10 days.

Now, the second window will be examined, Fig. 3.27

In this second window, the distance is higher than the *Distance to buy*, so the
algorithm stays out of the market.

This example shows how the algorithm takes the investment decisions. The
simulation is for the simple chromosome structure, but for the other two structures
the process is similar.

3.5 Conclusions

In this chapter the SAX representation method and several different chromosome
structures, to be used in the next chapter to validate this innovative approach, were
presented. From the description of the SAX-GA approach it is clear that this

symbolic form of time series representation will be more suitable to use in the GA, since the genes will contain a set of possible discrete symbols, rather than some double precision number, to generate the patterns to be discover. The fact of limiting the data representation will also limit the effect of overfitting, which usually exists on this type of algorithms, where it is necessary to do some training process, causing the method to tightly adjust to the training data resulting in a poor level of generalization.

References

1. D. Goldin, Kanellakis P, On similarity queries for time-series data: Constraint specification and implementation. *Proceeding First International Conference Prince Practice of Constraint Program.* (1995), pp. 137–153
2. J. Lin, E. Keogh, L. Wei, S. Lonardi, Experiencing SAX: A novel symbolic representation of time series. Data Min. Knowl. Discov. **15**(2), 107–144 (2007)
3. J. Lin, E. Keogh, S. Lonardi, P. Patel, Finding motifs in time series. Proceeding Eighth ACM SIGKDD International Conference on Knowledge Discovery and Data Mining Second Workshop on Temporal Data Mining, (2002), pp 53–68
4. J. Lin, E. Keogh, S. Lonardi, B. Chiu, A symbolic representation of time series, with implications for streaming algorithms. Proceeding Eighth ACM SIGMOD International Conference on Management of Data, Workshop on Research Issues in Data Mining and Knowledge Discovery, (2003), pp. 2–11. doi:10.1145/882082.882086
5. A. Canelas, R. Neves, N. Horta, A new SAX-GA methodology applied to investment strategies optimization. *Proceeding Fourteenth International Conference Genetic and* Evolution *Computing (GECCO' 12).* (2012), PP. 1055–1062. doi:10.1145/2330163.2330310
6. B.T. Zhang, J.J. Kim, Comparison of selection methods for evolutionary optimization. Int. J Evol. Optim. **2**(1), 55–70 (2000)
7. P. Parracho, R. Neves, N. Horta, Trading with optimized uptrend and downtrend pattern templates using a genetic algorithm kernel. *IEEE Congress on Evolutionary Computation 1895–1901*, (2011). doi:10.1109/CEC.2011.5949846

Chapter 4
Results

Abstract In this chapter the application created to test this new methodology is presented. Also, the results from the approaches described in the previous chapter are presented and compared. The several chromosome structures were tested in real market conditions, where in all transactions the costs were considered. In order to test this new approach of investment based on pattern discovery in financial time series, two major experiences were made. The first test is based on the discovery of patterns to invest *Long* presented like "SAX-GA Uptrend Pattern Discovery". The second test is a method that invests *Long* and *Short*, to do this it tries to discover patterns to enter and exit *Long* and another set of patterns to enter and exit *Short*, described later on "SAX-GA Multi-Chromosome Pattern Discovery".

Keywords Investment Strategies · Pattern Discovery · Genetic Algorithm · Buy and Hold

4.1 Application Structure

Based on the descriptions of the SAX-GA method of Chap. 3, an application in C++, to test the several approaches, was created [1]. Although the several different characteristics of individuals and evolution process of the populations, the program flowchart presented in Fig. 4.1 is applicable to all of them. The difference is in the modules that calculates/processes the distances and investment decisions that implements the rules defined on each methodology.

The application, as most of GA programs, could be divided in two major sections. The first section will be the training process where the parameters and patterns will be optimized and the second part where the best chromosome from the training section will be tested to prove the validity of the solution. Small

A. M. L. Canelas et al., *Investment Strategies Optimization Based on a SAX-GA Methodology*, SpringerBriefs in Computational Intelligence, DOI: 10.1007/978-3-642-33110-7_4, © The Author(s) 2013

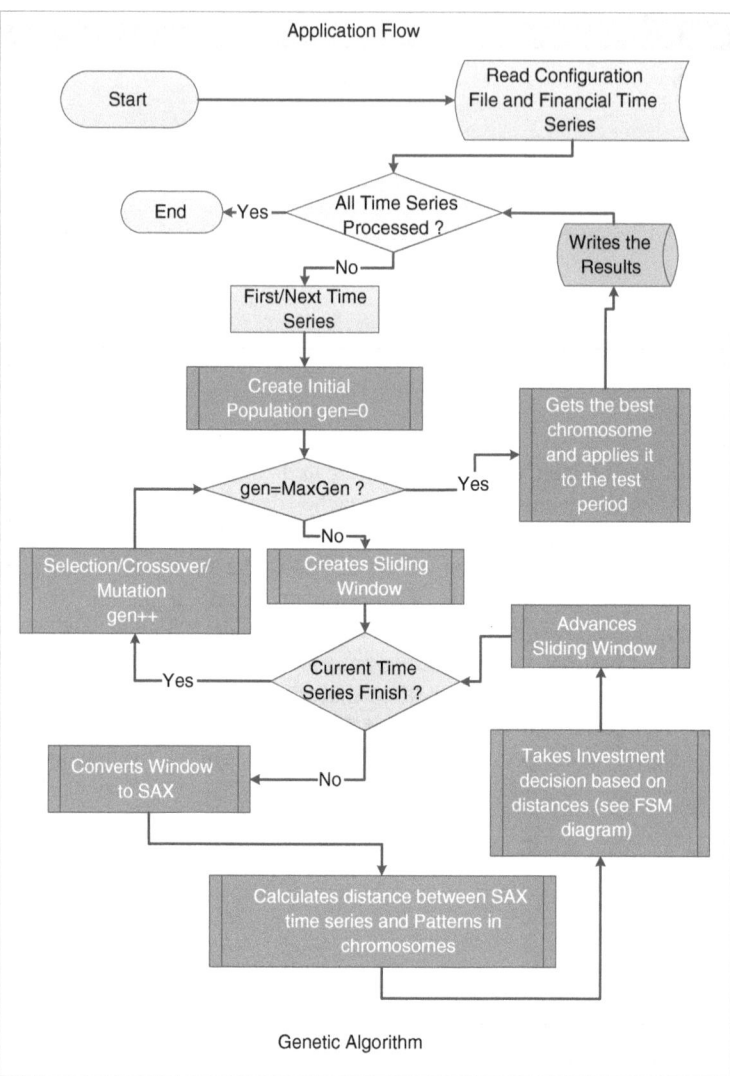

Fig. 4.1 Application flowchart

differences exists between the two sections, the evolutionary process of the population is the only difference, since in the test period only one individual exists, the best from the training period. That is why the test in the flowchart is presented by one sub-process box. In fact, this process includes the creation of a sliding window the creation of the SAX sequence, the distance calculation between the series and patterns and finally the investment decision, all these steps are identical to the training period.

Table 4.1 Tests made to validate 10 runs

Run	Average ROI (%)	Absolute ROI variation (%)
1	35.1805	–
2	39.3885	11.96
3	37.8213	3.98
4	38.5070	1.81
5	37.2306	3.31
6	37.7952	1.52
7	37.3621	1.15
8	37.8543	1.32
9	37.4147	1.16
10	37.4383	0.06
11	37.4233	0.04
12	37.4420	0.05

4.2 SAX-GA Results

To test the SAX-GA, a data set of 100 stocks was chosen from S&P500. Each asset has an historical price from 1998 onward. Thus, the stocks history prices range between January 1st of 1998 and April 21st of 2010, which was the last trading day when this testing period took place. The computing algorithm would then perform a training period with duration of approximately 7 years (January 1, 1998 to December 31, 2004), and a testing period with duration of 5 years and 4 months (between January 1, 2005 and April 21, 2010). There were no major criteria in the decision that involved the choosing of the 100 stocks from S&P500. So, volatility and other risk factors that could have been associated with the chosen financial assets were neglected. The test period was chosen to include the 2008 crash, in order to verify how the algorithm responds to crisis, even if the system is trained in non-crisis environment. Also for comparison purposes the raw data will be test by the genetic algorithm. The metric used to evaluate the methodologies performance is the ROI indicator, Eq. 4.1, and unless stated otherwise this indicator will be always calculated for the entire test period.

$$ROI(t) = \frac{P(t) - P(0)}{P(0)} = \frac{P(t)}{P(0)} - 1 \tag{4.1}$$

where,
$P(0)$ is the price at the initial time instant;
$P(t)$ is the price at the time instant t.

In all the methodologies tested 10 runs were used. This value was used after making several tests, where the average ROI value was evaluated after a new run was completed, and as can be seen by the Table 4.1 the ROI value appears to asymptotically converge to the value 37.4 % and after the tenth run the variation was around 0.06 %. The tests made with other stocks are quite similar validating the assumption that 10 runs are sufficient to consolidate the tests results.

Table 4.2 SAX-GA independent basic chromosome results

Point	SAX-GA ROI				Buy and hold ROI	
	Averaged 10 runs (%)		Best run (%)		In the period (%)	Annual (%)
	Period	Annual	Period	Annual		
P_1	3.6	0.67	18.5	3.25	0.32	0.06
P_2	−0.48	−0.09	17.8	3.14		
P_3	37.6	6.21	37.6	6.21		

4.2.1 SAX-GA Uptrend Pattern Discovery Results

In this section two different case studies were made. In the first one, the points identified in Sect. 3.1.4 and several other points were tested, using the basic chromosome structure presented in Fig. 3.13. The second tests the extended chromosome structure shown in Fig. 3.16, where the points are chosen by the algorithm.

4.2.1.1 Case Study I—Basic Chromosome Structure

In this case studies two tests are presented, the first analyzing all the stocks independently, the second trying to find the most important pattern for all the stocks.

For both tests, the population size was 500 individuals and the number of generations used in the stop criteria was 50. The tests were repeated ten times in order to consolidate results.

The range values for the genes in the chromosome are as follow:

- **Distance to Buy**—double between 0 and twice the word size
- **Distance to Sell**—double between 0 and twice the word size
- **Days to Sell**—integer between 1 and 100
- **Distance Measure Type**—0 or 1

Independent Pattern Discovery

In this test the stocks are consider individually and the method will try to identify patterns that occur on each financial time series. The program will treat each stock separately and discover patterns and investment rules. The SAX parameters used are the ones identified in Sect. 3.1.4, others combinations of SAX parameters were also tested in order to explore other areas of the solution space.

All the results are compared to the Buy & Hold investment strategy, which is widely used as reference, based on the efficiency of the markets [2].

The three points detected as regions of interest in the solution space, see Sect. 3.1.4, and were evaluated for the index S&P 500, which is an important indicator of the market behavior. The results of all runs and of the best one, on this financial asset are presented in Table 4.2.

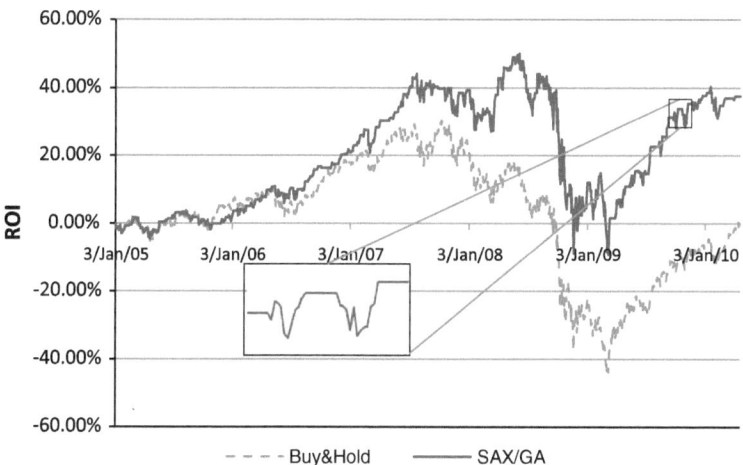

Fig. 4.2 S&P500 investment—B&H return vs. SAX-GA return

By the previous results, it is interesting to see that the combination of SAX parameters that give best results is for a pattern of only two word symbol. In all the runs for this point p_3, the patterns the GA identifies are descendent ones like 'FB', 'FA', 'EB' or 'DB', and the *Days to Sell* term is 11 for all the runs. So the algorithm tries to find minimum prices to buy and shortly sells them, after 11 days or if the stock price goes quickly apart from the pattern and this occurs when rapid up movement of the price manifests. As can be seen in Fig. 4.2, in the magnified area, where this fact is present. The problem is that this strategy only works well on bull markets. When the market is on a down movement, like in the 2008 period, the algorithm buys the stock because the pattern is present, but the market keeps losing, like is shown in Fig. 4.2, and the algorithm starts also to lose because the pattern keeps present and only leaves the position after the 11 days, to enter in the next day, since the down pattern keeps appearing. This strategy gives good results because the down period, compared to the rest of the testing period, is quite small.

In order to identify other strategies that can work well in bear and bull markets the rest of stocks were analyzed. It was also explored other areas of the solution space with different and more complex patterns. For example with a *Window Size* of 92, an *Alphabet Size* of 6 and a *Word Size* of 9, the GA found a pattern that gives good results in the "Alcoa Inc." stock, Fig. 4.3.

In Fig. 4.3 it is possible to observe how the algorithm avoids big drops, and saves the investment from falling in 2008. The pattern found and used by the SAX-GA is 'BDBCEFBDC' and could be represented by Fig. 4.4.

Although this pattern is similar to a head and shoulders, in fact this is not the case. The final section of the graphic (right side) should have a big drop, but instead is a drop that indicates a probable inversion on trend, it has smaller angle of inclination than the previous drops. So again, SAX-GA method tries to find

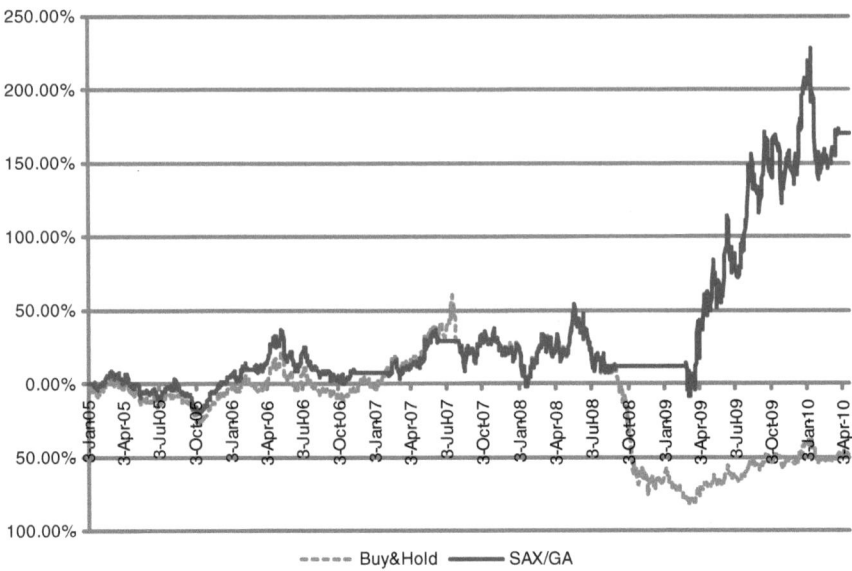

Fig. 4.3 Alcoa, Inc. B&H return vs. SAX-GA return

Fig. 4.4 Pattern
'BDBCEFBDC', found by
SAX-GA in the Alcoa
investment test

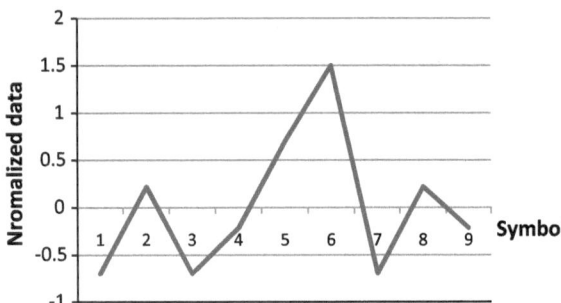

patterns that may occur in minimums to buy the stock and sells it when the price pattern goes away from this minimum pattern.

As was said before, other points were tested the results are presented in Table 4.3. Here are evaluated the average ROI of the investment strategies for the best chromosomes and best run of all the stocks, compared to the B&H. From Table 4.3 it is possible to verify that areas of SAX parameters, others than the detected in Sect. 3.1.4, are valid and prove to get better results. As expected the pattern search for a two symbol word gave the worst results on average.

Another study, concerning the measure types used to determine the degree of similitude between the time series and the pattern, was made. To find which distance measure is used with better result in the investment strategies were analyzed only the best runs of the GA, the results are presented in Table 4.4. In this table is possible to verify that the method chosen by the GA to find the best

Table 4.3 Average ROI for all stocks investment strategy

Sax parameters			SAX-GA ROI (%)		Buy and hold ROI (%)	
n	a	w	Period	Annual	Period	Annual
10	7	2	69.30	10.44	48.80	7.79
10	18	3	117.82	15.82		
17	13	4	115.18	15.56		
60	13	4	113.28	15.36		
92	6	9	120.49	16.09		
104	6	8	115.24	15.56		
128	7	8	111.43	15.17		
136	5	12	122.39	16.28		

Table 4.4 Distance measure type used by SAX-GA results

Measure type	Occurrences	
MINDIST (Eq. 3.5)	97	69.8 %
ALPHAB. DIST. (Eq. 3.7)	42	30.2 %

investment strategy was the distance measure usually defined by the SAX method, Eq. 3.5, which is a lower bounding of the Euclidian distance of the time series and the pattern. The sum of occurrences in the result table is bigger than 100, the number of stocks, because more than one run has the same result.

Best Pattern Discovery for all Stocks

As was said in the beginning of this Sect. 4.2.1.1, the first study treats the stocks separately, and tries to identify patterns and investment rules for each of them. In order to identify a global pattern and rule, it was feed to the GA all the stocks at once to the training period of the GA. The objective was to identify a pattern and rule that could be applied to any stock.

This test begins by the analysis of the alternative distance measure used by the GA to find the best solutions. The results are presented in Table 4.5. In this table is possible to verify that the measure select by the GA was, once again, the 'MINDIST', Eq. 3.5, which gives better results.

After the training period the best pattern found to invest on all the stock is in Fig. 4.5, the discovered pattern was 'ABCDAEBB'. This pattern provides the best returns on average for all the stocks, the SAX parameters associated with it are:

- Window Size = 128;
- Alphabet Size = 5;
- Word Size = 8.

For the discovered pattern, Fig. 4.5, a portfolio of all the stocks was made, and the investment strategy was applied to this set of stocks in the test period. The results are presented in Fig. 4.6. In this graphic is possible to see that the pattern takes profit by avoid the big dropdowns, in the more stable uptrend the results are

Table 4.5 Distance measure type used by SAX-GA best pattern results

Measure type	Occurrences	
MINDIST (Eq. 3.5)	94	77 %
ALPHAB. DIST. (Eq. 3.7)	28	23 %

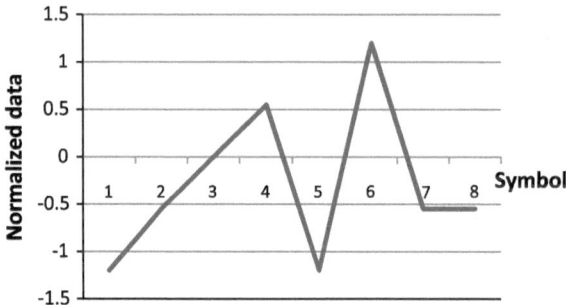

Fig. 4.5 Representation of the best global pattern found

Fig. 4.6 Investment returns of the portfolio

Table 4.6 Results of global investment strategy

Point	SAX-GA ROI		Buy and Hold ROI	
	Period (%)	Annual (%)	Period (%)	Annual (%)
P_1	51.44	8.15		
P_2	52.84	8.33	48.80	7,79
P_3	52.09	8.23		

below the B&H. This fact is explained by the fact that the algorithm will sell its position every 88 days, *Days to Sell* gene, and the transaction costs will reduce the profit.

Table 4.6 presents the average best results of this investment strategy for the points identified in Sect. 3.1.4.

The results from this test are not as good as the previous one, this was expected because in this global test a pattern that will suite to all the stocks is what the algorithm is trying to find. Although the lower ROI of this approach a better solution than B&H was also still found.

4.2.1.2 Case Study II: Extended Chromosome Structure

This case study evaluates the SAX-GA approach when the chromosome structure presented in Fig. 3.16 is applied. Based on this new structure several tests have been made, that results from the addition of the two SAX parameters, *Window Size* and *Word Size*, also changes in the investment strategy were applied. The tests were made changing the alphabet size, two values were tested. The first value used was an alphabet with 8 symbols size, the second value tested was a 12 symbols size. Those values were chosen because 8 was the upper limit of the study [3], and by the SAX parameter tests in Sect. 3.1.4. The 12 value was chosen because is the integer average of alphabet size parameter from the three points identified in Sect. 3.1.4.

The other parameter that was changed concerns to the investment strategy, in order to assess the relevance of the *Days to Sell* gene, it was added a control variable that allows enabling or disabling the evaluation of this gene in the investment strategy. This test will permit to understand if the discovered patterns have a time duration effect or if they are time independent.

For the tests the population size was 300 individuals and the number of generations used in the stop criteria was 50. The tests were repeated ten times in order to consolidate results.

The range values for the genes in the chromosome are as follow:

- Distance to Buy—double between 0 and twice the word size
- Distance to Sell—double between 0 and twice the word size
- Days to Sell—integer between 1 and 100
- Word Size—integer between 2 and half of the window size
- Window Size—integer between 4 and 120
- Distance Measure Type—0 or 1

In Table 4.7 the result from the parameters changes, previously discussed, are shown. From the two tests in alphabet size the small advantage goes to an alphabet size of 8 symbols. The results when the *Days to Sell* gene is enabled (1) or disable (0) show also a small distance between ROI's, to evaluate how the return is affected in each of the alphabet size by the *Days to Sell* gene another test was made and the results are also shown in Table 4.7.

From the previous table, is possible to conclude that when small alphabets are used, the *Days to Sell* gene effect is more important. This is because, rough patterns, less symbols, and could be detected more frequently than a finer detail pattern. So, the gene has an effect of limiting the transaction period, and as was

Table 4.7 Average ROI when parameters change

Alphabet size	'Days to Sell' gene	Average ROI in the period (%)
8	–	38.38
12	–	36.96
–	0	36.24
–	1	39.11
8	0	35.91
8	1	40.86
12	0	36.57
12	1	37.36

Fig. 4.7 Extended chromosome best investment strategy found by SAX-GA

seen before, the algorithm tries to find minimums and sells the stock few days later, considering the market in the test period this could avoid big losses.

The patterns found by the GA are very different from each other and on average the *Window Size* applied was 67 and the *Word Size* was 16. Those values are calculated over all the stocks and all runs. If only the best runs are evaluated, those parameters change to *Window Size* = 57 and *Word Size* will maintain in 16. From those results a constant was identified, the *Word Size* with 16 symbols, it is also possible to conclude from the data analysis that in both cases the compression factor found by representation was bigger than 3.5.

From the 100 stocks the best investment strategy was identified, based on the best return in absolute value from the B&H strategy. This best result was found in the Cliffs Natural Resources Inc. stock, in Fig. 4.7 this investment result is shown.

The pattern found by the GA in the previous example is presented in Fig. 4.8, this pattern is clearly identified as a bottom, so again the GA found a minimum to invest.

In Fig. 4.9 the average return from the best runs of the extended chromosome strategy is presented.

Fig. 4.8 Pattern found in extended chromosome best investment strategy

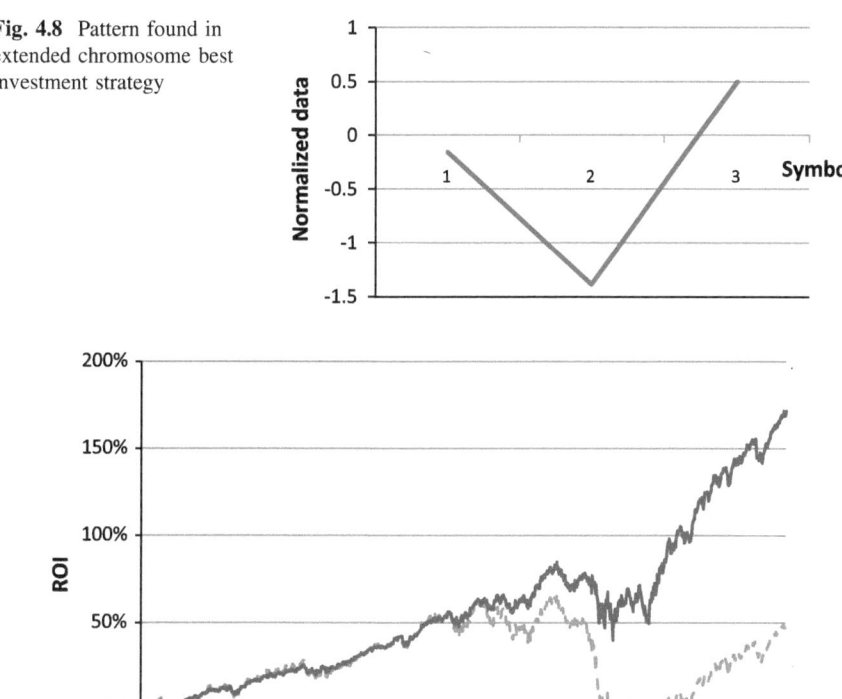

Fig. 4.9 Average return of extended chromosome SAX-GA vs. B&H

This graphic shows that on the first half, the SAX-GA is almost equal to the B&H, the method only proofs to beat the B&H when more instability becomes present and big drops began to appear. Although the SAX-GA was trained in a stable period in this test period where the markets are more unstable the method proof to behave well and it finds investment patterns that make sense. For instance the pattern presented in Fig. 4.10 that shows a double bottom and was found when the Constellation Energy Group stock was tested.

4.2.2 Multi-Chromosome SAX-GA Results

In order to test the multi-chromosome structure presented in Sect. 3.3 a set of experiences were made. Since this approach allows two investment strategies, *Long* and *Short*, several combinations have been tried with these strategies, in some cases it was allowed both, in others just one of them was allowed. Another

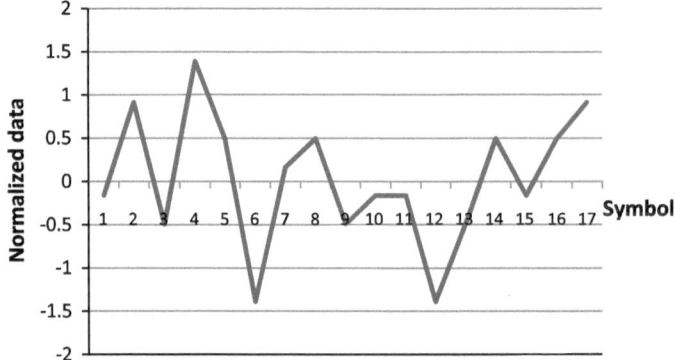

Fig. 4.10 Double bottom pattern found by extended chromosome SAX-GA method

Table 4.8 Multi-chromosome structure results—changing investment strategies and *Days to Sell/Buy* gene activation

'Days to Buy/ Sell' gene	Long strategy	Short strategy	Average ROI for all 10 runs (%)	Average ROI for the best run (%)
Enable	Enable	Enable	42.35	276.80
Enable	Enable	Disable	30.43	113.52
Enable	Disable	Enable	0.84	83.86
Disable	Enable	Enable	23.00	179.61
Disable	Enable	Disable	21.34	108.46
Disable	Disable	Enable	−9.69	52.74

parameter that was subject to changes was, like in the previous section, the activation or not of the *Days to Sell/Days to Buy* gene. For all the major tests the alphabet size was 8 symbols, but some tests were made using other values, to evaluate how this parameter affects the performance. Like in previous case studies 10 runs were made.

For the GA parameters a population of 200 individuals and 50 generations were used.

4.2.2.1 Investment Strategies combination

For this case study the *Days to Sell*/Days to *Buy* gene and investment strategies change according to the combinations presented in Table 4.8.

The range values for the genes in the chromosome of Fig. 3.18, are as follow:

- Distance to Enter/Exit—double between 0 and twice the word size
- Days to Sell/Buy—integer between 1 and twice the window size
- Word Size—integer between 2 and half of the window size
- Window Size—integer between 4 and 120

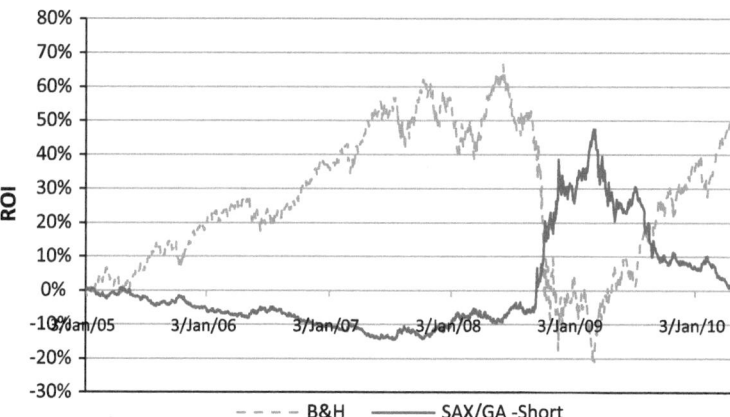

Fig. 4.11 Average ROI B&H vs. SAX-GA—short investment strategy

- Distance Measure Type—0 or 1

In Table 4.8 shows that the best results are obtained when the *Days to Sell/Buy* gene is activated. The best investment strategy is when the two trading methods are enabled, *Long* and *Short*, this is because in this case the algorithm could invest and profit in bear and bull market. The poor results when applying only *Short's*, are caused by the fact that the algorithm was trained on period that was indicated to *Long* investment strategy, so only when the algorithm gets to the 2008 crash it can make some profit, this can be seen in Fig. 4.11.

In Fig. 4.12 the return of the *Long* investment strategy is shown. In this figure is possible to observe that, on the left side of the graphic, the algorithm follows the Buy & Hold but with not so much profit, this is caused by the transaction cost that exists when the algorithm enter or exits the market. In the 2008 crash the algorithm does not drop as much as the Buy & Hold, in fact it recovers, but after again it goes apart from the Buy & Hold caused by the transaction costs. So the gains in this algorithm come from avoiding big drops.

In Fig. 4.13 the results from the combined *Long* and *Short* investment strategy is shown. In this graphic is possible to see the almost linear characteristic indicating that in all market types the algorithm can profit, sadly it cannot follow the Buy & Hold when the large gains occur, but on the other side when the market falls it keep gaining. The final result for the test period, in the present case, has a lower ROI of almost 6.5 % in relation to the Buy & Hold, again this is caused by the transaction costs.

Table 4.9 presents the parameters values found by the GA in this case study. The parameter *Days to Buy* is bigger than *Days to Sell* indicating that time to recover from the investment in *Short* takes more time than *Long*. The algorithm has two strategies to profit, in *Long* it invests in bottoms and sells few days after. In the *Short* case it sells in big drops and waits for the market to recover.

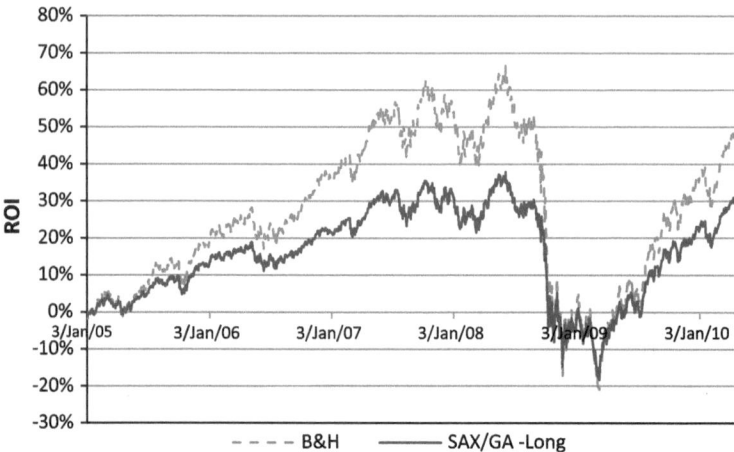

Fig. 4.12 Average ROI B&H vs. SAX-GA—long investment strategy

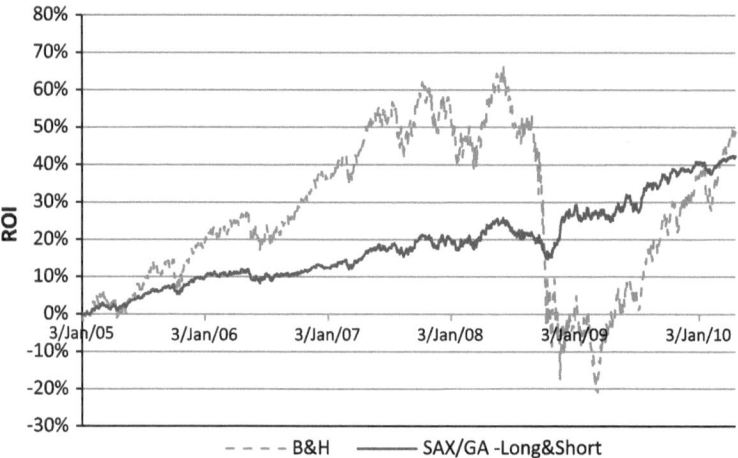

Fig. 4.13 Average ROI B&H vs. SAX-GA—Long&Short Investment Strategy

Table 4.9 SAX-GA average gene parameters values

Average runs	Window size long		Word size long		Days to sell	Window size short		Word size short		Days to buy
–	Enter	Exit	Enter	Exit	–	Enter	Exit	Enter	Exit	–
All	59.55	58.28	13.53	11,71	39.53	57.53	66.81	11.65	14.96	52.37
Best	46.12	59.76	12.24	14.21	39.64	61.52	73.93	13.22	18.19	75.77

In Fig. 4.14 is possible to see, for the Ball Corporation stock, the investment strategy found by the GA. In this graphic the *Long* and *Short* parts of transactions are in green and red respectively and in black are the periods where the algorithm

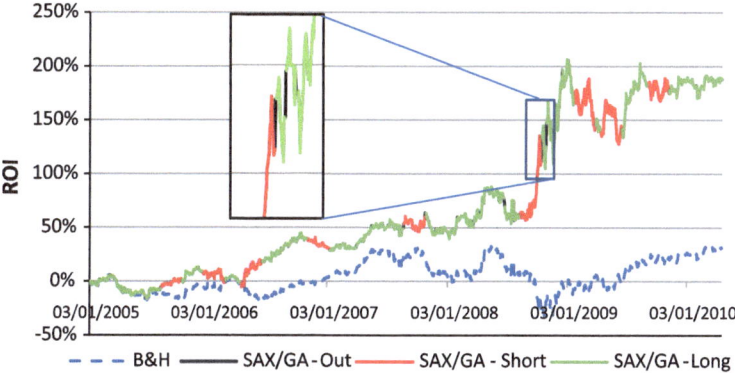

Fig. 4.14 Investment strategies *Long&Short* vs. B&H

is out of the market. In the highlighted area is possible to verify the small
dimension periods for the *Long* transactions and a longer period for the *Short*
transaction This is caused, as said before, by the fact that the GA buys the stock
and few days after sells it. The longer investment time for the *Short* strategy is
caused by the condition that usually precedes the bigger drops, where the price
stops growing and begins to go sideways, so this area is identified as belonging to a
period that should be used by the *Short* strategy creating longer periods for this
type of investment.

In Figs. 4.15 and 4.16 the patterns used by *Long* strategy in the previous
example are shown. The trend lines, in black, indicates that the GA buys in periods
where the price begins to grow, like bottoms, and sells on tops where the price
starts to drop or an inversion in trend begins to appear.

Figures 4.17 and 4.18 shows the patterns used by the SAX-GA *Short* strategy
when investing on the Ball Corporation stock, Fig. 4.14. In these images the trend
line, in thin black, shows that the GA makes a correct identification of patterns to
invest using a *Short* methodology.

4.2.2.2 SAX-GA Multi-Chromosome Alphabet Size Change Results

In order to evaluate the effect of changing the number of symbols in the alphabet
that represents the time series and patterns, a test using several values was made.
The new parameter values will be compared with the results from the previous
section, where 8 symbols were used.

In this case study, all the other parameters from the GA are the same as in the
previous test, the same population size, generations and all the range values for the
chromosome genes. The algorithm will invest using both *Long* and *Short* strategies
and the *Days to Sell/Buy* genes will be activated.

Fig. 4.15 Pattern to enter
Long—'CCBADAD
AHEACDGFAD'

Fig. 4.16 Pattern to exit
Long—'EAFAHEF'

Fig. 4.17 Pattern to enter
Short—'BFDAGGC
DHHADF'

Fig. 4.18 Pattern to exit
Short—'EEGGBFAED
CDAABCECCADEH'

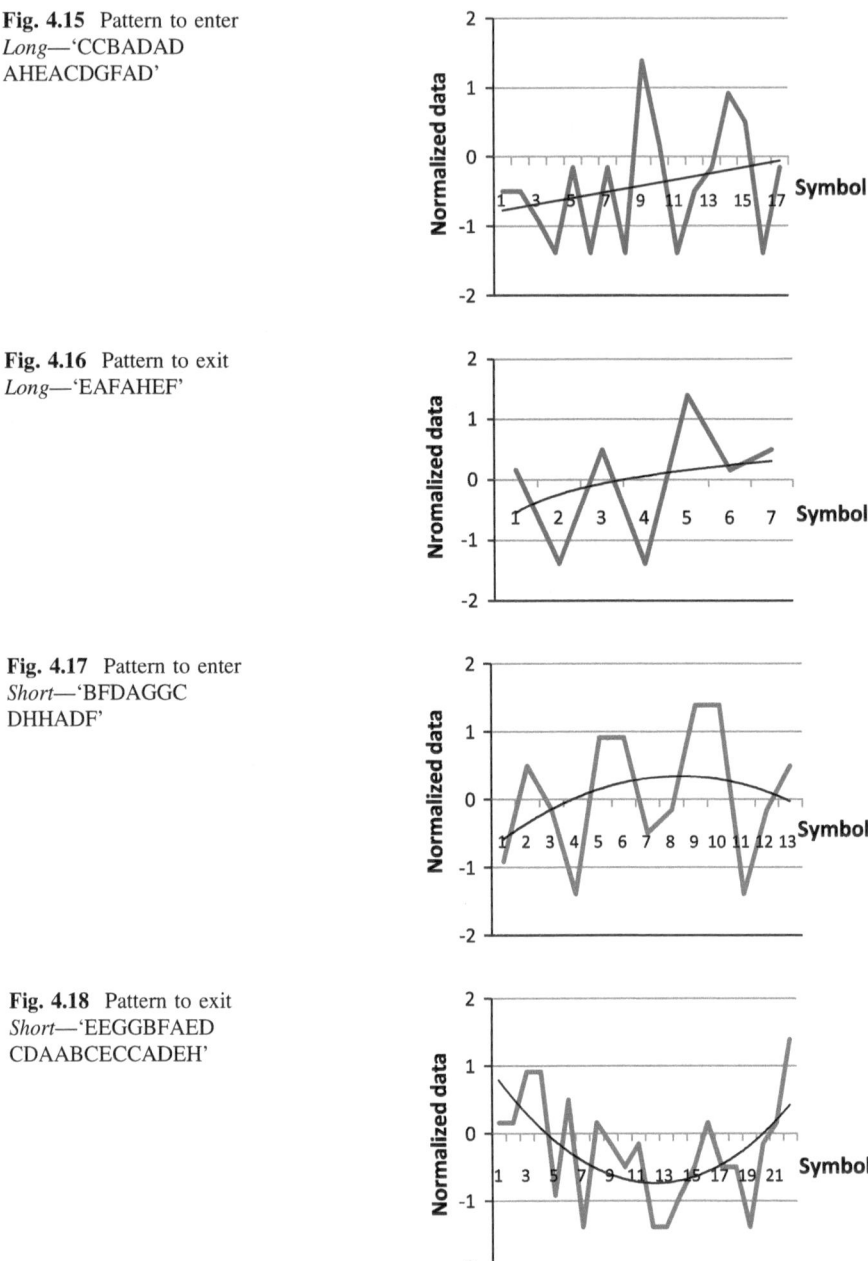

Table 4.10 shows the effect of changing the alphabet size parameter, this table shows that the best return occur for an alphabet size of 5 symbols, the second best is a 12 alphabet symbol.

Table 4.10 SAX-GA multi-chromosome results for alphabet size changes

Alphabet size	Average ROI for all 10 runs (%)	Average ROI for the best run
5	62.76	482.41
8	42.35	276.80
12	51.00	393.72
18	36.44	262.89

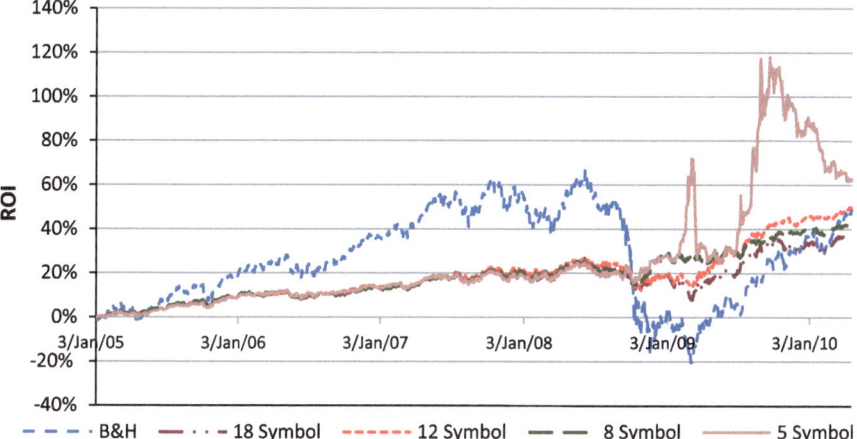

Fig. 4.19 SAX-GA Multi-Chromosome—Average ROI when changing alphabet size

To verify how the method profits according to the changes in the number of symbols in Figs. 4.19 and 4.20 are presented the result for the average ROI of all runs and the average ROI for the best runs respectively.

The previous figures show that the GA has large profits by avoiding the crash in 2008. These drops on prices produce patterns that usually are easier to find, since down movements are fast and relatively straight and are usually preceded by some level of resistance in the price. This factor makes it easy to be represented by a small symbol alphabet. The surprise here was the fact that an alphabet of 12 symbols produces so good results. So to try to understand why these values produce better result, than the other two cases, the parameters found by the GA were examined. One important value in the SAX representation is the dimensional reduction rate, defined by the relation between the *Window Size* and *Word Size*. This value is presented in Table 4.11.

If the values from the previous table were considered as points in a four dimensional space and the distance between them is calculated in order to identify clusters of similar relations, Table 4.12 shows the calculated distances between solutions.

From Table 4.12 the conclusion is that the alphabet size of 5 and 12 are closer than the other values, so the relations presented in these cases must be important to justify the better results in the test period. Although the alphabet size did not

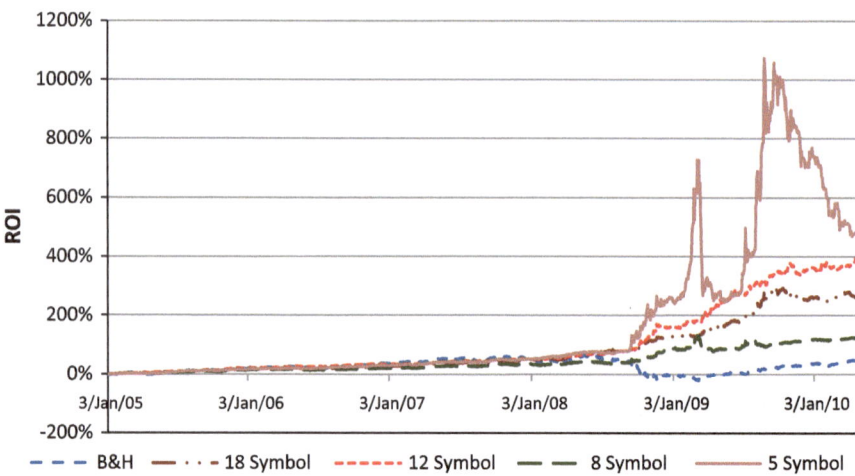

Fig. 4.20 SAX-GA Multi-Chromosome—Best runs when changing alphabet size

Table 4.11 SAX-GA multi-chromosome—average *Window Size/Word Size* results

Alphabet size	Relation between window and word size			
–	Enter long	Exit long	Enter short	Exit short
5	5.55	6.03	5.80	5.10
8	4.57	4.24	6.30	5.90
12	5.32	5.75	6.15	5.07
18	5.38	6.05	8.78	5.53

Table 4.12 Distance between relations points according to different alphabet size

Alphabet size	5	8	12
8	2.248222	–	–
12	0.504678	1.885206	–
18	3.015725	3.196795	2.687397

directly affect the *Window Size/Word Size* relation, it seems that leads the GA to a combination of values, that result on close parameters proximity between 5 and 12 symbols alphabet.

The proximity relation between those points is clearly valid for the data tested in this work, for other data probably some test must be made in order to identify the best alphabet size representation.

Table 4.13 Comparison results between single pattern detection methods

Method	Average return of 10 runs of all stocks (%)	Average return of best run of each stock (%)	B&H average ROI (%)
Basic chromosome	46.24	122.39	48.80
Extended chromosome	40.86	176.41	

Table 4.14 Comparison results between multi-chromosome structure and Buy and Hold

Method	Average return of 10 runs of all stocks (%)	Average return of best run of each stock (%)	B&H average ROI (%)
Multi-chromosome	62.76	482.41	48.80

4.3 Results Analysis

In this section the results from previous tests are compared and the new SAX-GA approach will be evaluated against the Buy & Hold approach.

In Table 4.13 a comparison between similar methods is presented. Those methods are similar because they all detect only one pattern to enter the market and exits when the pattern is no longer present.

The previous table shows that on average the best method is the "SAX-GA Basic Chromosome", but for the best runs is the extended chromosome approach. The poor result from the average extended chromosome method, was clearly caused by an extreme over fitting. In the training period, this method was able to achieve a ROI's as high as 60,000 %, so when the GA gets to test period it behaves worst than the basic chromosome. This is a serious problem when dealing with GA and if the number of parameters to be optimized is high and they are allowed to freely take values, this causes an additional effect of over fitting, since it will provide solutions more adjusted to the training period. This fact is the reason why the basic chromosome gets best result, because the SAX parameters are imposed and are not involved in the optimization process.

Table 4.14 shows the results for the Multi-Chromosome structure, when several patterns are being searched and two investment strategies are allowed.

This last table shows that the new SAX-GA approach gets better results than the Buy & Hold. This is because the SAX-GA is able to adapt and find new patterns that adjust to the market conditions.

4.4 Conclusions

The results obtained by the SAX-GA approach, prove that this method is capable of producing better results than Buy & Hold, when using the Multi-Chromosome structure. The ability of the GA to discover meaningful patterns and implement trading rules capable of using those patterns becomes clear when looking at the results and the discovered patterns. The patterns found by the SAX-GA methodology, are often known patterns, like the ones presented in Chap. 2. In several occasions, the patterns to enter *Long* are the same identified to exit *Short*, this fact makes all the sense, since this are opposite investment strategies. Many of the patterns discovered, by the Multi-Chromosome structure are difficult to identify and analyze, but looking at the trends defined by those patterns is possible to understand why the algorithm chooses them.

References

1. A. Canelas, R. Neves, N. Horta, A new SAX-GA methodology applied to investment strategies optimization in *Proceedings 14th International Conference Genetic and Evolution computation (GECCO' 12)*, (2012), pp. 1055–1062. doi:10.1145/2330163.2330310
2. P.L. Bernstein, A new look at the efficient market hypothesis. J Portf. Manag. **25**(2), 1–2 (1999)
3. J. Lin, E. Keogh, S. Lonardi, B. Chiu, A Symbolic representation of time series, with implications for streaming algorithms in *Proc. 8th ACM SIGMOD International Conference on Management of Data, Workshop on Research Issues in Data Mining and Knowledge Discovery*, (2003), pp. 2–11. doi:10.1145/882082.882086

Chapter 5
Conclusions and Future Work

Abstract This work presents an innovative method for discovering graphic pattern formations on financial data charts. The lack of this type of solutions, capable of discovering new patterns and adapt to new market conditions, is caused by the difficulty that this task represents, most of the approaches are parameter based relying on technical or fundamental indicators to optimize and generate trading rules, leaving out the important information that resides on the graphic price chart.

Keywords SAX-GA Strategy · Overfitting · Pattern Discovery

5.1 Conclusions

The implemented approach uses the Symbolic Aggregate approXimation method, as a representation for the financial time series, combined with a Genetic Algorithm optimization kernel to discover new patterns and implement trading rules capable of benefiting from the patterns recently discovered. The method was tested using data from the S&P500 index and used in real market conditions where the transaction cost were considered. In order to validate the approach taken, the test results were compared with the Buy & Hold strategy, the SAX-GA proved better than the B&H traditional strategy, Table 5.1. The other solutions found in the financial area, try to find known patterns and then apply some predefined decisions to invest, considering that the market nowadays changes very quickly and new patterns and opportunities are created, that kind of methods are inadequate to maximize investment results. The SAX-GA tries to find new patterns and adapt to new market conditions so it is able to get better earnings.

A. M. L. Canelas et al., *Investment Strategies Optimization Based on a SAX-GA Methodology*, SpringerBriefs in Computational Intelligence, DOI: 10.1007/978-3-642-33110-7_5, © The Author(s) 2013

Table 5.1 SAX-GA results versus Buy & Hold

Method	Average return of 10 runs of all stocks (%)	Average return of best run of each stock (%)	B&H average ROI (%)
Basic chromosome	46.24	122.39	48.80
Extended chromosome	40.86	176.41	
Multi-chromosome	62.76	482.41	

One of the major conclusions from the results was that the decision of joining SAX with the GA proved to be a good choice. The discretization used in the SAX representation method fits very well with the gene structure used by the GA. This combined approach was able to detect simple and complex patterns thanks to the adjustable structure of the chromosomes that adapts to the needs of the search. The potential of having structures that are dynamic and could adapt during the optimization process, was an important factor on the success of this work. Also, the possibility of having a multi-chromosome structure that allows to optimize and search for several patterns at once was an important characteristic. This kind of flexibility would be very difficult to implement on other approaches, like ANN or SVM. The results found proved a good ability of generalization even during the hard period of 2008, where a big crash hit the stock market. These results reveal that this approach has lots of potential and still has room to improvements as will be detailed on the future work section.

5.2 Future Work

A large problem present on this type of algorithms is the overfitting, and this work was no exception. In several runs was detected an excessive adjust of the parameters to the training data set, resulting on poor generalization capability during the test period. So, one of the projects for future work would be the reduction of this effect on the SAX-GA method, other changes are already also planned, which will be detailed next:

- A study of the optimization parameters in order to specify a tighter set of range values where these parameters could take values, in order to eliminate the freedom of choice by the GA to fine tune the parameters to the training data set needs with the objective of reducing the effect of overfitting.
- To improve the results and also reduce the overfitting effect a change on the training/testing process will be made, the present approach uses a large fixed training period followed by the testing period. To improve the method a sliding window strategy will be implemented. This will allow the testing period to be closer to the training reality.
- Studying the benefits that could come from the substitution of the basic SAX representation by new and recent evolutions of SAX, like the eSAX, that was created to work with financial time series.

- Another possibility of change in the representation of data is the creation of a SAX-GA solution where to each of the points that constitute a pattern a weight is associated, in order to define important points in the pattern, this solution tries to incorporate the benefits from the PIP method into the SAX-GA approach.